Coupled Phase-Locked Loops

Stability, Synchronization, Chaos and Communication with Chaos

WORLD SCIENTIFIC SERIES ON NONLINEAR SCIENCE

Editor: Leon O. Chua
University of California, Berkeley

*To view the complete list of the published volumes in the series, please visit:
http://www.worldscientific.com/series/wssnsa

WORLD SCIENTIFIC SERIES ON
NONLINEAR SCIENCE   Series A   Vol. 93

Series Editor: Leon O. Chua

Coupled Phase-Locked Loops

Stability, Synchronization, Chaos and Communication with Chaos

Valery V. Matrosov
Vladimir D. Shalfeev

Nizhny Novgorod University, Russia

World Scientific

NEW JERSEY · LONDON · SINGAPORE · BEIJING · SHANGHAI · HONG KONG · TAIPEI · CHENNAI · TOKYO

Published by

World Scientific Publishing Co. Pte. Ltd.
5 Toh Tuck Link, Singapore 596224
USA office: 27 Warren Street, Suite 401-402, Hackensack, NJ 07601
UK office: 57 Shelton Street, Covent Garden, London WC2H 9HE

Library of Congress Cataloging-in-Publication Data
Names: Matrosov, Valery V., 1960–   author. | Shalfeev, Vladimir D., author.
Title: Coupled phase locked loops : stability, synchronization, chaos and communication with chaos /
    Valery V. Matrosov, Vladimir D. Shalfeev, Nizhny Novgorod University, Russia.
Other titles: World Scientific series on nonlinear science. Series A,
    Monographs and treatises ; v. 93.
Description: Singapore ; Hackensack, NJ : World Scientific Publishing Co. Pte. Ltd., [2018] |
    Series: World Scientific series on nonlinear science. Series A, Monographs and treatises ; vol. 93 |
    Includes bibliographical references and index.
Identifiers: LCCN 2018019174| ISBN 9789813271944 (hardcover ; alk. paper) |
    ISBN 9813271949 (hardcover ; alk. paper)
Subjects: LCSH: Phase-locked loops--Mathematics. | Nonlinear oscillators--Mathematics. |
    Chaotic behavior in systems.
Classification: LCC QA612.76 .M38 2018 | DDC 515/.39--dc23
LC record available at https://lccn.loc.gov/2018019174

British Library Cataloguing-in-Publication Data
A catalogue record for this book is available from the British Library.

For any available supplementary material, please visit
https://www.worldscientific.com/worldscibooks/10.1142/11033#t=suppl

Typeset by Stallion Press
Email: enquiries@stallionpress.com

Contents

Preface

Nonlinear science is now at the stage of intensive development. Modern technical, biological, socioeconomic systems and processes are extremely complex. Nevertheless, in spite of this complexity, the behavior of these systems over relatively large periods of time is determined by relatively simple dynamical models and patterns. In the early stages of development of nonlinear dynamics, these patterns were associated with stable solutions such as equilibrium states, limit cycles, tori, and later, strange attractors. Nowadays, transient motions and metastable states get in the spotlight of the dynamics of complex systems, as their signatures are broadly found in neurodynamics and socioeconomic systems. The study of the processes of evolution of complex systems and their understanding largely relies on concepts of competition and cooperation (synchronization). These processes are likely to prove cornerstones of the dynamics of complex systems.

The main approaches to the study of the dynamics of complex systems are now associated with models of collective dynamics of networks or ensembles formed by interacting dynamic elements. Examples include a large series of works on collective dynamics of ensembles of interacting Chua's circuits [1], neural networks [2], collective dynamics of socioeconomic networks [3], collective dynamics of networks of coupled phase-controlled oscillators [4], etc. Investigations of the collective dynamics of such networks largely employ numerical methods, since the applicability of analytical and qualitative methods of nonlinear dynamics is severely restricted due to the high dimension of the phase space. Therefore, studying the simplest models of networks — ensembles with a small number of interacting elements — becomes of particular interest. Such models allow for making use of the whole spectrum of analytical, qualitative and numerical methods of nonlinear dynamics. The book is devoted to the investigation of a kind of such systems, namely, small ensembles of coupled phase-controlled oscillators. We address both traditional issues like synchronization relevant for applications in radio communications, radiolocation, energy, etc., and non-traditional issues of excitation of chaotic oscillations and their possible application in advanced communication systems.

The authors are grateful to V. S. Afraimovich, A. V. Gaponov-Grekhov, A. S. Dmitriev, M. V. Kapranov, V. P. Ponomarenko, M. I. Rabinovich, N. F. Rulkov, D. I. Trubetskov, D. V. Kasatkin, K. G. Mishagin, M. V. Ivan-chenko, M. V. Shalfeeva, L. O. Chua, G. Chen, J. Kurths for their valuable help and their discussions.

Chapter 1

Introduction: Dynamical Chaos and Information Communication

The discovery of dynamical chaos oscillations of deterministic origin possessing the properties of random processes, such as continuous spectrum, finite correlation time, high sensitivity to perturbations, unpredictable behavior on large time intervals — the brightest event in nonlinear science in the recent decades. The considered oscillations are self-oscillatory motions of nonlinear systems and their properties are fully determined by the properties of the dynamical system. The understanding that complex noise-like oscillations (dynamical chaos) may arise in nonlinear systems at zero random action changed the traditional conception of oscillatory and wave phenomena dramatically. It became clear that chaotic oscillations of one or another degree of chaoticity rather than simple periodic oscillations are typical for the majority of physical, chemical, biological and other natural systems. The discovery of dynamical chaos aroused great interest in the scientific community and has been a challenging field of research since then.

The pioneering works on dynamical chaos were published at the beginning of the 1960s and initiated theoretical and experimental studies of this phenomenon in many laboratories worldwide. The basic concepts related to dynamical chaos were formulated by the beginning of the 1990s [4–11].

Investigation of the principal properties of dynamical chaos stimulated an interest in the application of this phenomenon in engineering systems that would use the features of dynamical chaos. One of the promising trends was to use chaos in communication systems. Dynamical chaos possesses many attractive properties that may be useful for information transmission [12]:

- a possibility of producing complex oscillations by means of devices with a simple structure;
- realization of a great variety of chaotic modes in one device;
- controlling chaotic modes by small variations of system parameters;
- large information capacity;
- diversity of methods of introducing information signal into a chaotic one;
- higher modulation velocity as compared to regular signals;
- novel methods of multiplexing;
- secure communication.

The trend of using dynamical chaos for communications started at the beginning of the 1990s when quite a few chaotic oscillators were created, such as inertial nonlinearity oscillator [13], ring self-excited oscillator [9], tunnel diode oscillator, Chua's circuit [7], and the phenomena of chaotic synchronization [14–18] and chaotic synchronous response [19] were discovered.

Beginning in 1992, some methods of information communication using chaotic signals were proposed: chaotic masking, chaos shift keying, nonlinear mixing, etc. They demonstrated a possibility of using chaos for data communication, thus creating the background for the development of the new trend in communication systems.

Early experiments on information communication [20–25] confirmed that it is possible, in principle, to transmit information using chaos. However, development of this trend faced severe problems. Firstly, communication circuits based on self-synchronization of chaotic oscillations are very sensitive to distortions in the channel, noises and incomplete identity of transmitter and receiver parameters. Secondly, in the proposed circuits, chaos was used as undercarrying oscillations modulating a highfrequency carrier. As a result, the attractive feature of chaos, that is, its broadband that was intended to ensure high-speed communication of signals with large data base, was lost. Consequently, further development of chaotic communication systems demanded solutions for a number of topical problems, which included elaboration of high-efficiency generators of chaos operating in a straightforward manner in a wide frequency range covering high and ultrahigh frequencies, and creation of methods that would ensure stable synchronization of chaotic oscillations.

It seems promising to employ phase and frequency systems as a solution to these problems [26]. Information communication devices in which regular signals are used are known to broadly employ phase-locked loops (PLL) and frequency locked loops (FLL). These systems were initially developed for solving the problems of synchronization, frequency stabilization, controlling the frequency and phase of radio oscillations, filtration, demodulation, signal formation and processing, and some other problems. High accuracy, reliability, noise stability, controllability, capability of operating at high and ultrahigh frequencies, and technological effectiveness made these systems an essential part of nearly all communication systems. Naturally, the features enumerated above make such systems attractive for creation of novel communication systems using chaotic signals instead of the traditional regular ones.

At present, the phase and frequency systems for regular signals have a well-developed theory [27–32], whereas the application of PLL and FLL for information communication using chaotic signals still lacks theoretical basis. The results available in the theory of dynamical behavior of such systems in asynchronous modes are insufficient for explaining numerous phenomena occurring outside the domains of existence and stability of the synchronous regime (including domains of existence of self-modulation modes), for defining dynamical characteristics of chaotic signals generated by these systems, as well as for goal-oriented development of communication systems using chaotic modes of PLL and FLL.

A specific feature of the considered class of systems is the presence of phase or frequency control circuits, allowing the frequency of the controlled oscillators to be stabilized relative to the regular reference signals in a wide range of initial deviation of frequency. However, outside the synchronization domain, these circuits provide a rich potential to excite various self-modulation oscillations, including the chaotic ones. It should be emphasized that the chaotic signals formed at the output of controlled PLL oscillators may be transmitted to the communication channel immediately after formation, without additional transformations, which is an undoubted merit of these systems. Another advantage of the considered systems is their ready integration into an ensemble by means of different couplings between PLL and FLL systems. The new properties arising as a result of such integration expand the functional potential of the systems, both as traditional frequency controlled generators of periodic signals and generators of chaotic signals. In practice, the considered systems are integrated in an ensemble when it is necessary to meet conflicting requirements to different characteristics: pull-in band, filtering properties, speed, cycle skip probability, etc. [28, 30–32]. There are also problems when several systems should be unified in an ensemble [4], for instance, optimal reception and assessing parameters of complex signals. As for new applications, particularly for chaotic communication systems, of major interest are complex chaotic self-modulation oscillations that are realized in such ensembles. Note that dynamical properties of self-oscillatory systems are determined not only by parameters of the systems but also by the structure and force of coupling between the systems, thus allowing coupling parameters to be used as control ones.

Note that the considered ensembles of coupled voltage controlled oscillators are a variety of multielement self-oscillatory systems that currently attract the attention not only of physicists but also of biologists, chemists, economists, and so on [4]. The nonlinear phenomena of collective dynamics demonstrated by such systems (synchronization processes, self-oscillatory regular and chaotic modes) are, firstly, important for ascertaining the basic regularities of dynamical behavior of coupled frequency and phase-controlled oscillators and, secondly, they may be useful for investigation of other objects (multielement-phased arrays, Josephson contacts, power circuits, spatio-temporal processing systems, etc.).

The book consists of eight chapters. The first chapter is introduction. The second chapter is concerned with the investigation of the nonlinear dynamics of a typical model of PLL system with second-order filter described by a third-order differential equation defined in cylindrical phase space. The solutions to the model are analyzed using the mathematical apparatus of phase space. The correspondence between attractors of the mathematical model and real dynamical modes of the PLL system is established. These dynamical modes may be classified into three groups: synchronous, quasisynchronous and asynchronous. The quasisynchronous PLL modes are especially interesting for the formation of chaotic oscillations with angular modulation and stabilized carrier frequency. In the study of the nonlinear dynamics of the PLL model, primary attention is focused on the investigation of methods to achieve quasisynchronous modes and their chaotization processes, analysis of the properties of chaotic modes, and study of the structure of regions of existence of quasisynchronous modes in the space of parameters.

Chapters 3–5 address the problems of nonlinear dynamics of small ensembles of PLLs. Models of two and three cascade-coupled PLLs are studied in Chapters 3 and 4, and features of behavior of an ensemble of two PLLs coupled in parallel are considered in Chapter 5. Dynamical properties of the ensembles are investigated by analyzing trajectories in the phase spaces of the corresponding mathematical models described by systems of ordinary differential equations. The models of the phase systems are defined in cylindrical phase spaces containing several angular coordinates. This significantly complicates the study of the dynamics of phase-controlled oscillators as the considered models have a greater variety of motions than those defined in Cartesian coordinates. It is shown that the integration of phase-controlled oscillators in an ensemble results in changes in the characteristics of the dynamical modes of partial systems and in the appearance of new modes that are not typical of partial systems. Scenarios of the onset of self-oscillatory modes, mechanisms of their chaotization, and domains of existence of chaotic oscillations in the space of parameters are investigated. The properties of chaotic oscillations are analyzed as a function of parameters of phase systems and coupling parameters. It is shown that the properties of the generated oscillations may be controlled quite efficiently by small changes in coupling parameters. Results of physical experiments with an ensemble of two cascade-coupled phase systems are presented. These experiments confirm the results of the theoretical studies and indicate that PLL ensembles are effective generators of chaotically modulated oscillations in a wide region of parameter space.

Problems concerned with synchronization of chaotically modulated oscillations of two unidirectionally coupled systems, both single and belonging to an ensemble, are addressed in Chapter 6. For synchronization of chaotic oscillations of phase-controlled oscillators with non-identical but close parameters, it is proposed to use the principle of automatic tuning (synchronization). It is demonstrated that by following this principle it is possible to synchronize chaotically modulated oscillations to a rather high accuracy and to broaden existence domains of the mode of chaotic synchronization in the space of parameters for both single-loop oscillators and oscillators combined in an ensemble.

Results of computer simulations of information communication using dynamical chaos are presented in Chapter 7. The experimental results lead to a conclusion that it is promising to construct new communication systems using dynamical chaos. Chapter 8 is the conclusion.

A few comments on the adopted terminology: The main topic of this book is collective dynamics of coupled phase-locked loops (PLL). In the literature, such systems are also frequently referred to as phase synchronized systems (PSS). Sometimes, they are called phase-controlled systems or simply phase systems (PS). The terms PLL, PSS and PS are used in the text as synonyms.

The term *synchronization of chaotic oscillations* is rather ambiguous in the literature and greatly differs from the term for regular periodic oscillations. To avoid ambiguity, we give some comments on the use of this term in Chapter 5 that is devoted to the synchronization effects.

Chapter 2

Nonlinear Dynamics
of the Phase System

2.1. Mathematical Models

The continuous system of phase-locked loops (PLL) is a typical ring of the automatic frequency control generator. Figure 2.1 (see [28]) shows the structural scheme of the continuous system of PLL. The basic elements of the ring are as follows: the controlled generator (G) (or voltage controlled oscillator), the phase discriminator (PD), the low frequency filter (F) and the control element (CE). The scheme functions according to the following principle. The periodical oscillations from the output of the generator G with the current phase θ_1 are compared at the phase discriminator to the oscillations of the reference signal with the current phase θ_0. As a result, a signal depending on the phase difference $\varphi = \theta_0 - \theta_1$ is formed at the output of PD. The signal further passes through the filter and is transmitted to the controlled element, which changes the frequency of the control generator in accordance with the frequency of the reference signal. The mathematical model of the PLL system (also referred to as the basic one) may be presented by the following equation [28, 32]:

$$\frac{p\varphi}{\Omega} + K(p)F(\varphi) = \gamma, \qquad (2.1)$$

where $p \equiv d/dt$ is the differentiation operator, Ω the maximum frequency detuning (which can be compensated for by the control circuit), $\gamma = \Omega_H/\Omega$ (where Ω_H is the initial oscillation frequency detuning), $K(p)$ the filter transfer ratio in the operator format and $F(\varphi)$ the standard characteristic of the PD. The nonlinear features

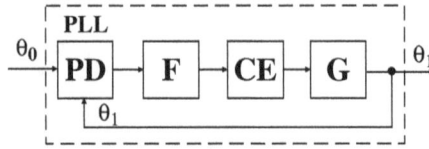

Fig. 2.1　A phase-locked loop.

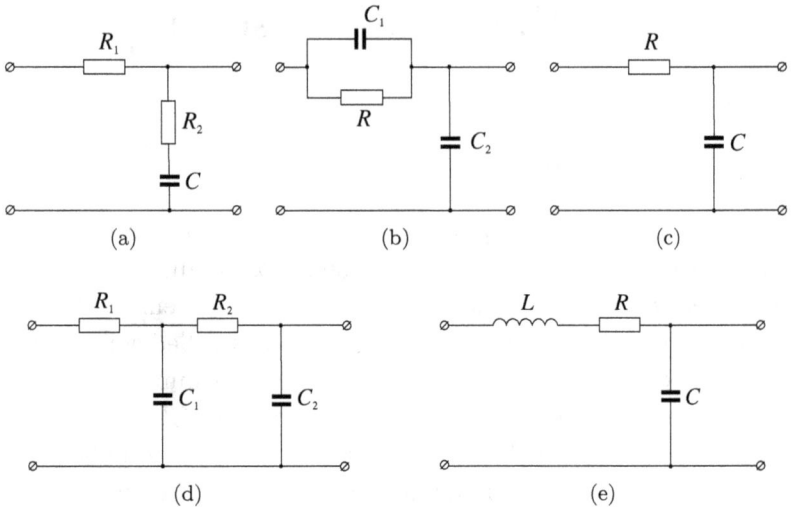

Fig. 2.2　The filter schemes are as follows: the first-order filters are proportionally integrating (a, b) and integrating (c); the second-order filters are two-section RC-filter (d) and the single mesh RLC-filter (e).

of Eq. (2.1) are determined by the nonlinear characteristics of the PD and the inertial ones depend on the filter in the control circuit. The choice of filter is based on providing the desired qualities of the dynamic processes. For any particular filter with the transfer ratio $K(p)$, it is possible to move from the symbolic representation of the model (2.1) to the PLL model by an ordinary differential equation; the order of which is determined by the filter type $K(p)$.

Figure 2.2 shows the schemes of the most frequently used filters. Figure 2.2 shows two schemes of the first-order filter of $[1/1]$ type (proportionally integrating). The filter has the operator transfer ratio of the following type: $K(p) = (1 + nTp)/(1 + Tp)$, where

$n = R_2/(R_1 + R_2)$, $T = (R_1 + R_2)C$ for the scheme in Fig. 2.2(a); $n = C_1/(C_1 + C_2)$, $T = R(C_1 + C_2)$ for the scheme in Fig. 2.2(b). If $n = 0$, this filter is an integrating RC filter (Fig. 2.2(c)) with the onset time $T = RC$.

The simplest second-order filter of the $[0/2]$ type with the transfer ratio $K(p) = (1 + a_1p + a_2p^2)^{-1}$ is either a two-section RC filter (Fig. 2.2(d)), or an RLC filter (Fig. 2.2(e)). In the first case, $a_1 = R_1C_1 + R_2C_2 + R_1C_2$, $a_2 = R_1C_1R_2C_2$, and in the second case, $a_1 = d/\omega_0$, $a_2 = \omega_0^{-2}$, where $\omega_0 = (LC)^{-1/2}$ is the filter border frequency and $d = R(L/C)^{-1/2}$ the filter attenuation.

If the proportionally integrating filter from (2.1) is used in the circuit, we obtain the following mathematical model:

$$\frac{d\varphi}{d\tau} = y, \quad \varepsilon\frac{dy}{d\tau} = \gamma - F(\varphi) - [1 + n\varepsilon F'(\varphi)]y, \qquad (2.2)$$

which is determined on the two-dimensional phase cylinder $V = \{\varphi \pmod{2\pi}, y\}$, where $\tau = \Omega t$ is the non-dimensional time, $\varepsilon = \Omega T$ and n are non-dimensional parameters of the low frequency filter.

Assuming that there is the second-order filter with the transfer ratio $K(p) = (1 + n_1a_1p + n_2a_2p^2)/(1 + a_1p + a_2p^2)$ in the circuit, we can obtain the following mathematical model from (2.1):

$$\frac{d\varphi}{d\tau} = y, \quad \frac{dy}{d\tau} = z,$$

$$\mu\frac{dz}{d\tau} = \gamma - F(\varphi) - [1 + n_1\varepsilon F'(\varphi)]y - n_2\mu y^2 F''(\varphi) \qquad (2.3)$$
$$- [\varepsilon + n_2\mu F'(\varphi)]z,$$

which is determined on the three-dimensional cylindrical phase space $U = \{\varphi \pmod{2\pi}, y, z\}$, where $\varepsilon = \Omega a_1$, $\mu = \Omega^2 a_2$, $0 \le n_1 < 1$, $0 \le n_2 < 1$ are the non-dimensional control circuit parameters.

2.2. Dynamical Modes and Characteristics

Nonlinear PLL can operate in various stationary modes [28, 31]. Attractors of various types can represent such modes in the phase spaces of the mathematical models. The PLL system with the more complicated filter is likely to demonstrate the broader range of dynamic modes. Thus, we will present the interpretation of the modes for PLL with reference to the model (2.3) [33].

The synchronization mode of the generator G with the reference signal, when the frequencies of the reference and the controlled generators are equal, the slow change in the parameters determine the frequencies that are fully addressed by the operation of the PLL system. The stable equilibrium O_1 (Fig. 2.3(a)) responds to the synchronization mode in the phase space of the mathematical models.

The quasisynchronization mode, when the oscillation at the PLL output is regularly or chaotically angle modulated near an average frequency, is stabilized by the reference signal. Regular (Fig. 2.3(b)) or chaotic (Fig. 2.3(c)) oscillation (limited by the coordinate φ) attractors correspond to this mode on the phase space.

It should be noted that there may be such attractors in the phase space of the model (2.3), which are characterized by the amplitude by the coordinate φ greater than 2π, but the number of rotations at φ coordinate on 2π is compensated by the number of rotations on -2π (Figs. 2.3(d) and 2.3(e)). Due to the limitation of the rotations of phase trajectories at the φ coordinate, such attractors are also oscillatory. The oscillatory attractor with the amplitude on φ, exceeding 2π, will be further referred to as the oscillatory attractors with phase rotation, and the modes corresponding to such attractors will be referred to as the quasisynchronous modes with phase rotation. In this case, we can observe the shifts of the oscillation phase of the control generator with respect to the reference signal on $\pm 2\pi$, while the average frequency of such oscillations is stabilized by the reference signal.

The beat mode is the one where the difference of the controlled and reference signals increases unlimitedly. These modes in the phase space correspond to the rotatory (Figs. 2.3(f) and 2.3(g)) and the oscillatory–rotatory (Figs. 2.3(h) and 2.3(i)) attractors (on the analogy with the pendulum motion). The rotatory and the oscillatory–rotatory attractions differ from each other by the fact that the φ coordinate on the rotatory attractors increases continuously while on the oscillatory–rotatory attractors the increase in the phase difference process contains the oscillatory stages. Beat modes, as well

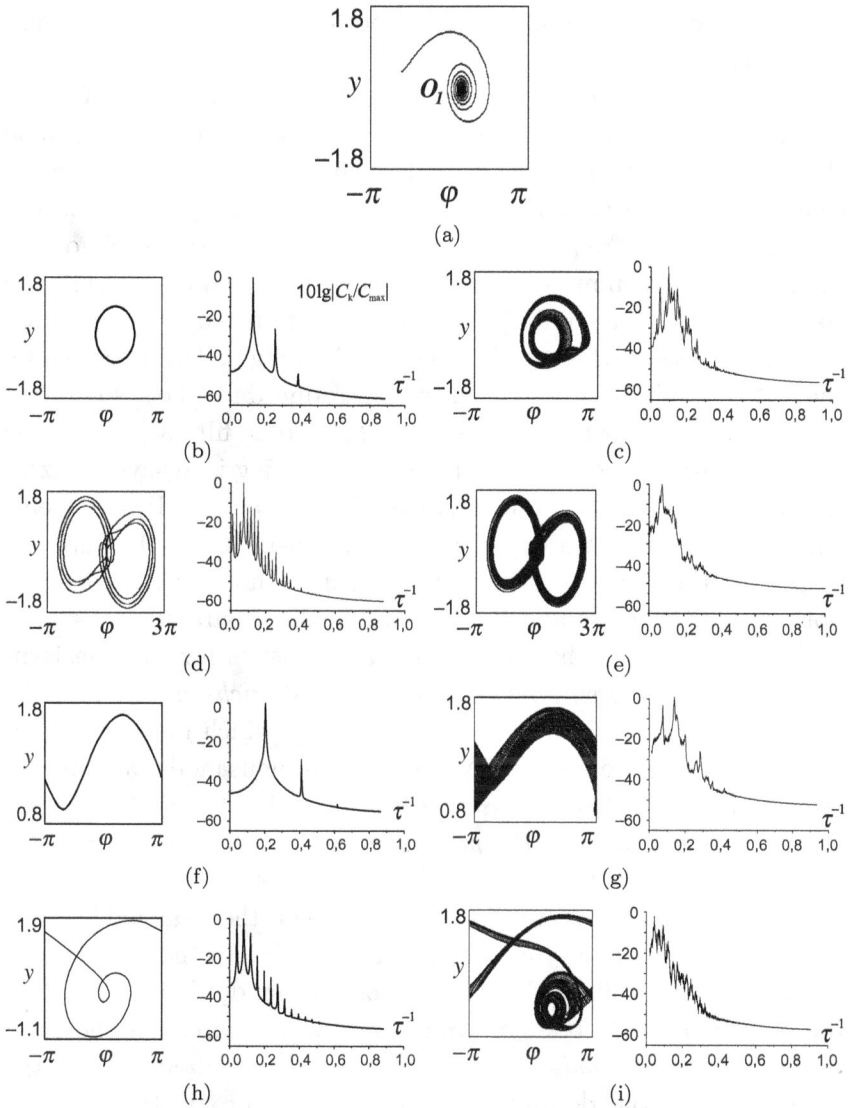

Fig. 2.3   The examples of (φ, y)-projection of the attractors and the corresponding spectra of the model (2.3), characterizing the basic dynamical modes of the PLL system.

as quasisynchronous modes, can be both regular (Fig. 2.3(h)) and chaotic (Fig. 2.3(i)).

The dynamic modes are inseparably connected with the dynamic characteristics of PLL. The synchronization mode is considered to be the basic dynamic mode. This mode holds in case of γ variations as characterized by the concept *PLL pull-out frequency band*, i.e., the band of the initial frequency detuning γ, where the process of holding the synchronization mode occurs.[1] It should be noted that usually the concept of pull-out band of the synchronization mode is connected with an increase of the initial frequency detuning γ. This is caused by the fact that initially the concept was introduced for the system with simple dynamics (PLL systems with first-order filters), where the synchronization mode at zero frequency detuning is always realized. There may be no synchronization mode at $\gamma = 0$ in PLL systems with the filters of a higher order, while it exists at $\gamma \neq 0$. Thus, the search for the borders of the pull-out range should be carried out both with an increase and the decrease in the γ parameter.

Let us introduce the analogous characteristics for the quasisynchronous mode: *the pull-out range of the quasisynchronous mode*, i.e., the range of the initial frequency detuning γ, which makes the quasisynchronous mode possible. As quasisynchronous modes are divided into regular and chaotic ones, we can consider *the pull-out range of the regular mode* and *the pull-out range of the chaotic quasisynchronous mode* separately.

The transmission state of the system when the beat mode eventually passes into the synchronization mode due to the γ variations is called the pull-in to the synchronization mode. The concept of *the pull-in range to the synchronization mode* represents the state of the system. *The pull-in range to the synchronization mode* is the range of the initial detuning γ, where the synchronization mode is established under any initial conditions. Let us analyze this concept

---

[1]The dynamic PLL modes are, as a rule, multiparameter, so a whole region of parameters correspond to the synchronization mode within the space of model parameter. Thus, alongside with the concept as *pull-out band of the synchronization mode*, we can also use an equivalent but a more general concept of *the pull-out region of the synchronization mode*, i.e., the multitude of all parameter values makes the synchronization mode possible. This note is quite general and refers to all concepts characterized by a range due to possible changes of the initial frequency detuning that preserves a dynamic mode.

with the reference to the quasisynchronization mode. Let us assume that *the pull-in range to the chaotic (regular) quasisynchronization mode* is the spectrum of the initial detuning γ, which makes the chaotic (regular) quasisynchronization mode possible.

Thus, besides the well-known and widely-used synchronization mode of the PLL generator oscillations, other less-known and less frequently used modes can be realized in PLL by the reference signal. The diversity of the dynamic modes of the PLL system proves that the system has big functional capabilities as the generator of different types of oscillations (both regular and chaotic). In order to use these system features in practice we should know the values and the mutual arrangements in the parameter spaces of the modes' existence regions; we should also be aware of the influence of the system parameters on the generated oscillation and their existence regions. These issues will be considered in the following sections.

2.3. The Phase System with the First-Order Filter

Equation (2.2) for PLL purposes was considered by Kapranov [34], Belyustina [35], Gubar, Bautin and others. Belyustina and Belykh gave a complete and detailed study of equation (2.2) for the general function $F(\varphi)$ in [36]. These results are shown in Fig. 2.4 as the (ε, γ) parameters plane partition and corresponding coarse phase portraits.

If the parameter values belong to the region D_1, there are two states of equilibrium on the phase cylinder V: globally stable state $O_1(\varphi = \varphi_1^*, y = 0)$ and the saddle one $O_2(\varphi = \varphi_2^*, y = 0)$, where $\varphi_2^* = \pi - \varphi_1^*$, $\varphi_1^* = \arcsin \gamma$. This is the region of the lock-on to the synchronous mode. The rotatory limit cycle L, which determines the regular beat mode, emerges at the transmission from region D_1 to region D_2 as a result of the bifurcation of the separatrix loop which encircles the phase cylinder itself. The transmission from region D_1 to region D_3 is accompanied by the saddle-node bifurcation,[2] giving birth to stable L and non-stable Γ limit cycles on the phase cylinder V.

---

[2]The saddle-node bifurcation of the limit cycle corresponds to the case when one of the multipliers of this cycle is equal to +1.

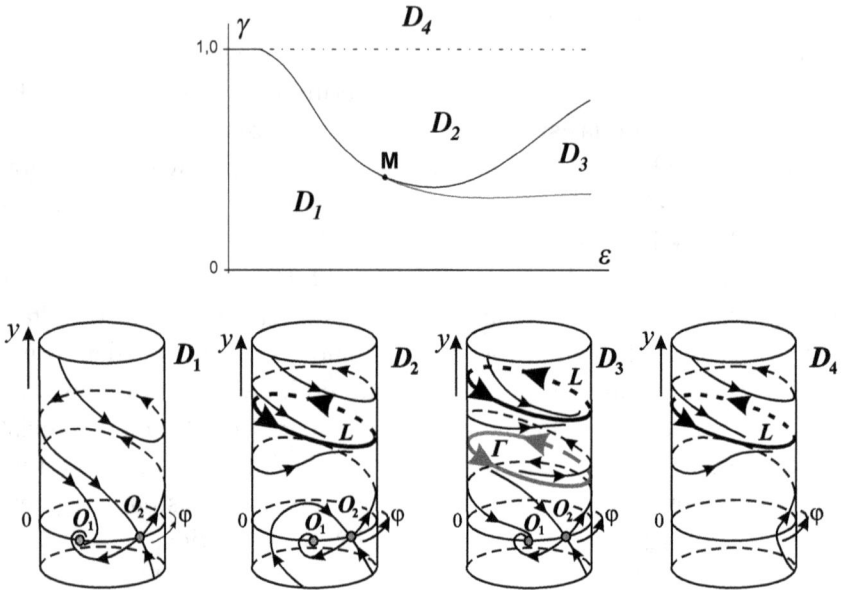

Fig. 2.4   The parametric and the phase portraits of the PLL model with the proportionally integrating filter in the control circuit.

While passing from region D_3 to region D_2, the cycle Γ gets into the separatrix loop of the saddle O_2. There are two attractors in regions D_2 and D_3: O_1 and L; consequently, both synchronous and asynchronous modes can be realized in these regions depending on the initial conditions. The dash-and-dot line $\gamma = 1$ is the border of the synchronous mode pull-out range. When $\gamma = 1$, the stable state of equilibrium O_1, responsible for the synchronization mode disappears, merging with the saddle state of equilibrium O_2. A part of the line $\gamma = 1$, imposing on the axis $\varepsilon = 0$, responds to the separatrix loop of the saddle node. When the parameters belong to region D_4 in the phase space system (2.2), there is a single attractor L, i.e., the beat mode is realized in PLL for any initial values. Thus, the border of the PLL system pull-out range with the proportionally integrating filter is the line $\gamma = 1$, and the border of the pull-in range to the synchronous mode on the left of the point M is the bifurcation curve of the separatrix loop and on the right of M is the bifurcation curve of the double limit cycle. The point M responds to the saddle value

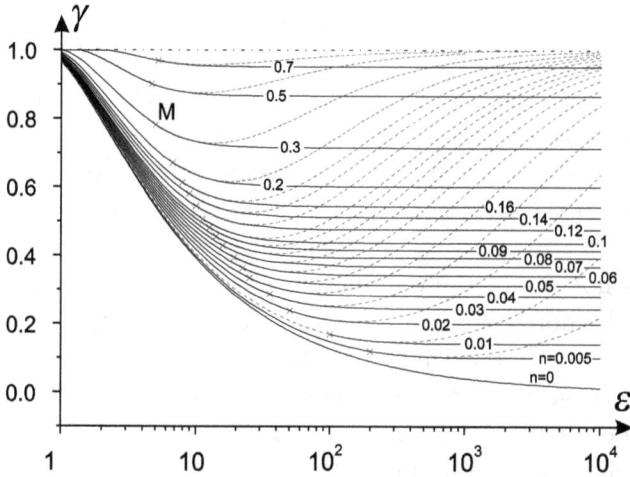

Fig. 2.5 The numerical estimation of the pull-in and pull-out ranges of PLL system with the proportionally integrating filter for different values of the parameter n.

reducing to zero. When there is an integrating filter in the control PLL system (the case $n = 0$), the border of the pull-in range to the synchronous mode is the bifurcation curve of the separatrix loop (Fig. 2.5).

It should be noted that even if a complete qualitative research of the dynamic system is available in phase space, the quantitative data is usually defined by computational numerical methods. Figure 2.5 shows the results of the pull-in region in the synchronous mode of the PLL system with the first-order filter [37]. The dotted lines in the figure continue the bifurcation curves of the separatrix loop, which are not included in the border of the pull-in range as they are responsible for generating the unstable limit cycle Γ from the separatrix loop.

2.4. The Phase System with the Second-Order Filter

Let us consider the dynamics of the PLL model with the second-order filter starting with the case $\mu \ll 1$. If $\mu \ll \max(\varepsilon, 1)$, the system (2.3) is a dynamic one with small parameters at the derivative being $dz/d\tau$. The full movement in the phase space U is divided [38] into "rapid" and "slow" movements. The surface U_0 of slow movements is defined by the equation $\gamma - F(\varphi) - [1 + nF'(\varphi)]y - \varepsilon z = 0$ and is stable in relation to the rapid movements. The equations of the slow movements on the surface U_0 coincide with Eq. (2.2). Thus, the dynamics of PLL system with the second-order filter if $\mu \ll 1$ is equivalent to the dynamics of the PLL system with the first-order filter.

If μ is not small, the dynamic processes in the system (2.3) are significantly more complex. New stationary modes appear, which are not typical for the two-dimensional model (2.2), in particular it is true about chaotic processes. Due to the significant nonlinearity and high dimensionality of the model (2.3), it was researched by numerical modeling based on the methods by the theory of oscillations and the bifurcation theory. The results of the research of the system (2.3), obtained on the basis of various assumptions and by different methods [39], were widely used. This information primarily concerns the bifurcation surfaces, which separate the space of the system (2.3) parameters into the regions where the system is globally asymptomatically stable and contains the cycles of the first and the second types and also has a lot of saddle cycles [39]. One of the first papers that considered the chaotic attractors was [33]. It stated the existence of oscillatory, rotatory and oscillatory–rotatory chaotic attractors for the system (2.3) in case of $F(\varphi) = \sin \varphi$, $n_1 = n_2 = 0$ by numerical modeling. The ranges of their existence on the plane of the parameters (μ, γ) were selected at fixed ε. In [40–44] that research continued. The works [45–50] are devoted to the dynamics of various PLL models.

According to [39], the model (2.3) possesses an unlimited number of bifurcations, so the complete bifurcation analysis is impossible. The numeric research described below was aimed at selecting with in the model parameters the regions with qualitatively different dynamic behavior of the system and studying the evolution of the selected regions if the model parameters changed. The analysis of the parameter space was carried out in two stages. At the first stage, the role of the parameter filters and the initial frequency detuning were analyzed for fixed characteristics of the PD by creating parametrical portraits in various sections. At the second stage, the changes of the selected structures in case of changing characteristics of PD were analyzed.

2.4.1. *The system with the filter [0/2] in case of zero initial frequency detuning*

The influence of the filter parameters on the dynamic modes can be studied with the use of the PLL model sinusoidal characteristics of the PD when $\gamma = 0, n_1 = n_2 = 0$. In this case, the system (2.3) is symmetrical regarding the change $\Pi_1 : (\varphi, y, z) \to (-\varphi, -y, -z)$. The possible modes of the system (2.3) behavior can be obtained by the bifurcation diagram $\{\mu, \varepsilon\}$ in Fig. 2.6. Line I is defined by the equation $\mu = \varepsilon$. It corresponds to the change in the stability of the equilibrium state $O_1(\varphi = 0, y = 0, z = 0)$ and the generation of the oscillatory limit cycle L_0. Line II corresponds to the soft separation from the cycle L_0 of the pair of asymmetric oscillatory cycles L_{01} and L_{02}. In this case cycle L_0 becomes unstable. Line III limits from below the region of the cycle existence L_{01} and L_{02}. Line IV

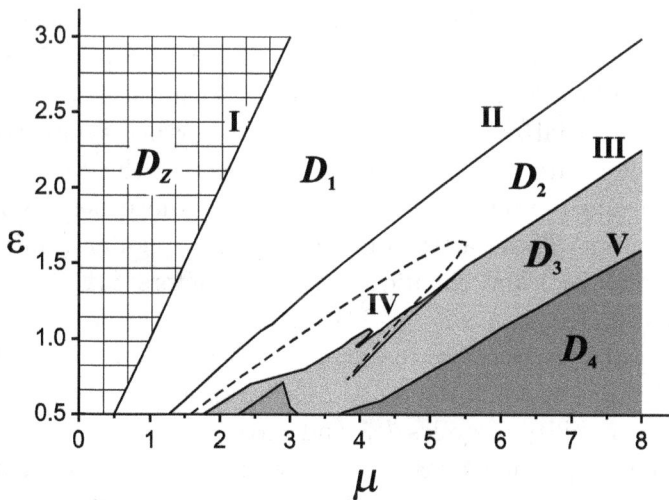

Fig. 2.6 The bifurcation diagram of the dynamic modes of model (2.3) if $\gamma = 0$, $n_1 = n_2 = 0$.

corresponds to the generation in the phase space of the system (2.3) of oscillatory attractors with phase rotation. When $\mu > 5$ this line coincides with curve III. When line V crosses with lessening ε the oscillatory attractors with phase rotation change into the oscillatory–rotatory ones.

The bifurcation curves shown in Fig. 2.6 allow us to select the ranges of the dynamic modes for the PLL system discussed here:

$D_Z = \{\varepsilon > \mu\}$ — the pull-in range in the synchronization mode. The only attractor of the system (2.3) in this range is the equilibrium state $O_1(\varphi_1^*, 0, 0)$.

D_1 — the range where a single quasisynchronous mode with regular modulation exists is located between lines I and II. The difference of frequencies and phase oscillations with parameter values from this range change with regard to zero value. The only oscillatory limit cycle L_0 (Fig. 2.7(a)), which is invariant with regard to the transformation Π_1 exists in this range.

D_2 — the range with bistable quasisynchronous PLL system behavior is located between lines II and III. The PLL system generates quasisynchronous oscillations with parameter values from this region. The phase differences for these oscillations change with regard to a certain average value $\tilde{\varphi} = \varphi_{01}$ or $\tilde{\varphi} = -\varphi_{01}$. Which of these two values $\tilde{\varphi}$ should be set in the system depends on the initial conditions. There is a pair of asymmetric (the attractors are not invariant with respect to the change in Π_1, this change transfers one attractor into the other) oscillatory attractors. These attractors can be both regular (limit cycles L_{01} and L_{02}) and chaotic (strange attractors SA_{01} and SA_{02}) [Figs. 2.7(b) and 2.7(c)]. The attractors become chaotic as a result of a series of bifurcation of the period doubling cycle. The dotted line crossing the range D_2 corresponds to the first change in the stability of cycles L_{01} and L_{02}.

D_3 — the range where quasisynchronous modes with the phase rotation exist is located between lines IV and V. Regular and chaotic oscillatory attractors with phase rotation are also found in this region. (Figs. 2.7(d–f)).

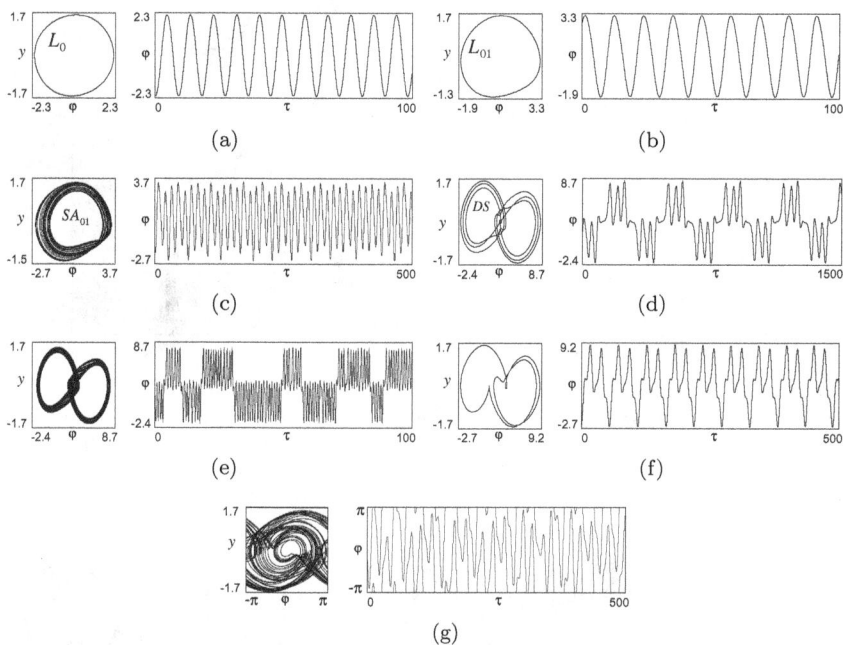

Fig. 2.7   The projections of the phase portraits of the attractors and the dependence $\varphi(\tau)$ of the system (2.3) if $\gamma = 0$, $\varepsilon = 1$, (a) $\mu = 2$; (b) 3; (c) 3.5; (d) 3.88; (e) 4.0; (f) 4.98; (g) 6.0.

D_4 — below line V there is the range of the asynchronous chaotic modes. In the phase space of the system (2.3), there is a chaotic oscillatory–rotatory attractor (Fig. 2.7(g)).

In the diagram $\{\mu, \varepsilon\}$, the range of the parameters with various types of system (2.3) movements indicates the development of the dynamic process. Let us consider the development of the processes in the system, which takes place with the increase in μ when the value $\varepsilon = 1$ is fixed. The evolution of the self-oscillatory mode in this case is illustrated by a single parametric bifurcation diagram $\{\mu, z\}$ of the Poincaré map in Fig. 2.8 and the projection of the phase portraits of the attractors in Fig. 2.9.

The evolution of the self-oscillatory modes of the model (2.3) with the increased μ starts with $\mu > 1$ ($\mu \in D_1$), when as a result of changing the stability of the equilibrium state O_1 a stable limit cycle L_0

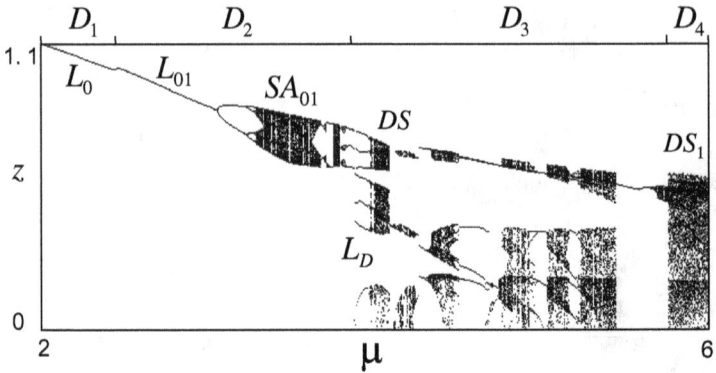

Fig. 2.8   Bifurcation diagram $\{\mu, z\}$ of self-oscillatory modes of the system with symmetry when $\varepsilon = 1$.

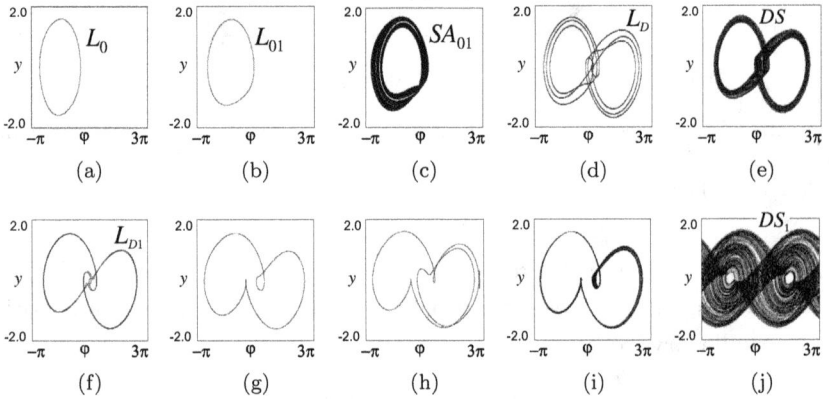

Fig. 2.9   Projections of the phase portraits of the attractors of the system with symmetry when $\varepsilon = 1$, (a) $\mu = 2$; (b) 3; (c) 3.5; (d) 3.88; (e) 4; (f) 4.12; (g) 4.7; (h) 4.98; (i) 4.745; (j) 6.

(Fig. 2.9(a)) softly emerges around it. The stable cycle L_0 exists in the interval $1 < \mu < 2.4418$. When $\mu = 2.4418$ ($\mu \in D_2$), the cycle L_0 becomes unstable and a pair of asymmetric cycles L_{01} and L_{02} appears in its neighborhood. As μ increases the periods, these cycles are doubled, thus, generating a pair of chaotic oscillatory attractors SA_{01} and SA_{02}. The projections (φ, y) of the phase portraits of the attractors L_{01} and SA_{01} are, respectively, shown in Figs. 2.9(b) and 2.9(c). It should be noted that the bifurcation diagram $\{\mu, z\}$ depicts the evolution of only one asymmetric attractor L_{01}. The branch of the curve corresponding to the evolution of the attractor L_{02} can be obtained from the branch of cycle L_{01} by changing z for $-z$, and the phase portraits of attractors L_{02} and SA_{02} can be obtained from the portraits in Figs. 2.9(b) and 2.9(c) by changing y, φ for $-y, -\varphi$. The chaotic attractors SA_{01} and SA_{02} exist in the interval $3.3 < \mu < 3.88$. The region of existence of chaotic attractors is typically intermitted by "windows" of multiturn limit cycles that transform to chaotic attractors again with a slight change in μ.

For $\mu = 3.88$ ($\mu \in D_3$), an oscillatory limit cycle with phase turns L_D (Fig. 2.9(d)) is realized in the phase space of the system (2.3). The image point on L_D makes three turns around the equilibrium state $O_1(0, 0, 0)$ and then again makes three turns around the equilibrium state $O_1'(2\pi, 0, 0)$ and again returns to the turns around $O_1(0, 0, 0)$. The cycle L_D is symmetrical. When μ increases on the basis of cycle L_D, there appears a symmetrical chaotic oscillatory attractor with phase rotation DS (Fig. 2.9(e)), which collapses at $\mu = 4.08$ and the system (2.3) passes to one of the oscillatory cycles L_{01} or L_{02}. With further increase of μ, cycles L_{01} and L_{02} transform to chaotic attractors SA_{01} and SA_{02}, which collapse at $\mu = 4.12$ and the system again returns to the symmetrical oscillatory cycle with phase rotation L_{D1} (Fig. 2.9(f)). Cycle L_{D1} unlike cycle L_D makes only two turns around the equilibrium state O_1.

Further development of dynamical processes occurs in the following scenario: *The symmetrical cycle is replaced by the symmetrical chaotic attractor that breaks down giving rise to a pair of asymmetrical cycles which further transform to asymmetrical chaotic attractors which, in turn, break down to form a symmetrical cycle.* This process is repeated up to $\mu = 5.74891$ ($\mu \in D_4$), when, as a result of formation of a homoclinic structure, the chaotic symmetrical oscillatory attractor with phase rotation transforms to a chaotic oscillatory–rotatory attractor DS_1. Examples of phase portraits of oscillatory attractors with phase rotation are shown in Figs. 2.9(d–f): regular symmetrical attractor (Figs. 2.9(d) and 2.9(f)), chaotic symmetrical

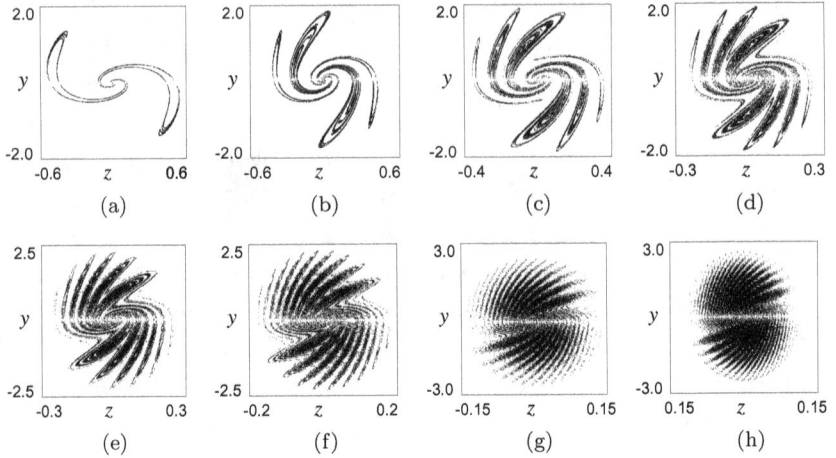

Fig. 2.10   Patterns of Poincaré mapping generated by trajectories of the attractor DS_1 for $\gamma = 0, \varepsilon = 1$, (a) $\mu = 6$; (b) 15; (c) 25; (d) 50; (e) 75; (f) 150; (g) 300; (h) 500.

attractor (Fig. 2.9(e)), regular asymmetrical attractor (Figs. 2.9(g) and 2.9(h)), chaotic asymmetrical attractor (Fig. 2.9(i)). The attractor DS_1, whose phase portrait is shown in Fig. 2.9(j), is conserved with a further increase of μ and its evolution with increasing μ is depicted by the corresponding patterns of Poincaré maps in Fig. 2.10.

Thus, the results presented above indicate that, for $\gamma = 0$, the system (2.3) can have attractors of different complexity. Appearance of oscillatory modes is caused by the loss of stability of equilibrium state and their further development proceeds, as a rule, by the following scenario: *a symmetrical oscillatory cycle, a pair of asymmetrical oscillatory attractors, an oscillatory attractor with phase rotation, and an oscillatory–rotatory chaotic attractor.* Chaotization of oscillatory attractors occurs through a period doubling bifurcation and is observed at small values of $\varepsilon < 1.5$ and $\mu < 5$, and the region of existence of oscillatory–rotatory chaotic attractors is adjacent to the straight line $\varepsilon = 0$. It is interesting to elucidate the influence of parameter γ on the modes that are set in the system, on their evolution, and regions of existence.

2.4.2. *The influence of the non-zero initial frequency detuning*

The influence of the parameter γ, which is accountable for the asymmetry in the system (2.3), is reflected in the parametrical portraits $\{\mu, \gamma\}$ and $\{\varepsilon, \gamma\}$, obtained for the sinusoidal characteristics of the PD $F(\varphi) = \sin \varphi$ and $n_1 = n_2 = 0$. The system (2.3) is invariant with respect to the transformation $\Pi_2 : (\gamma, \varphi, y, z) \rightarrow (-\gamma, -\varphi, -y, -z)$, that is why it is sufficient to consider the values $\gamma \geq 0$. The obtained curves and the regions restricted by the curves are given in the appendix in Fig. 2.11.

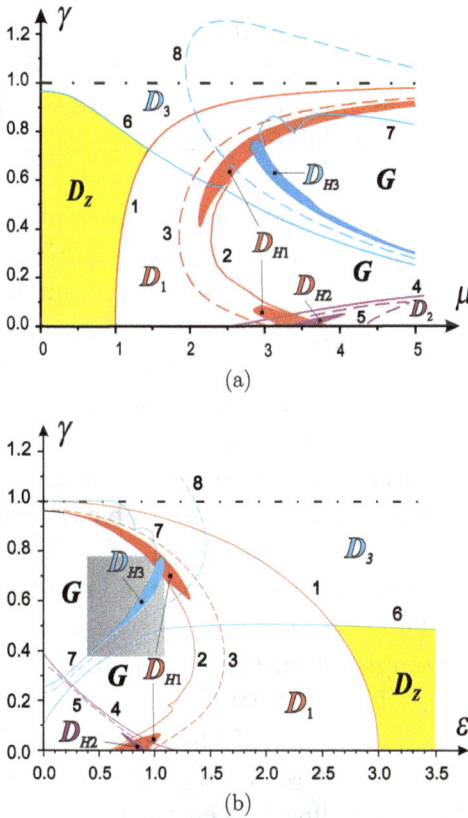

Fig. 2.11 The bifurcation diagram of the dynamic modes of model (2.3) if $n_1 = n_2 = 0$, (a) $F(\varphi) = \sin \varphi$, $\varepsilon = 1$; (b) $F(\varphi) = \sin \varphi$, $\mu = 3$.

The dot-dash straight line $\gamma = 1$ is the boundary of the region of existence of the equilibrium states of system (2.3). For $\gamma < 1$, two equilibrium states $O_1(\varphi^* = \arcsin\gamma, \ y^* = 0, \ z^* = 0)$ — that is a node or a focus, and $O_2(\pi - \varphi^*, y^*, z^*)$ — that is a saddle or a saddle-focus, exist in phase space U. The equilibrium state O_1 loses its stability through Andronov–Hopf bifurcation on the curve $\gamma = \gamma_s(\mu, \varepsilon, n_1, n_2)$ (line 1 in Fig. 2.11), which satisfies the equation $(\varepsilon + n_2\mu\sqrt{1 - \gamma_s^2})\,(1 + n_1\varepsilon) - \mu\sqrt{1 - \gamma_s^2} = 0$. The region D_U, restricted by the straight line $\gamma = 1$ and the curve γ_s, is the one holding the synchronization mode.

In region D_1 between lines 1 and 2, an oscillatory attractor around the equilibrium state O_1 exists in the phase space of system (2.3). The line 1, which restricts the region D_1 from above and on the left in Fig. 2.11 (or on the right in Fig. 2.11(b)) corresponds in this case to the soft change in the equilibrium state stability O_1 and the generation of a stable oscillatory limit cycle L_{01} (Fig. 2.12). With increasing distance from line 1 inside the region D_1, the cycle L_{01} may undergo a series of period doubling bifurcations and give rise to an oscillatory chaotic attractor S_{01} (Fig. 2.12(b)). The oscillatory chaotic attractor S_{01} exists for the parameter values of region D_{H1}. Line 2 bounding region D_1 on the right in Fig. 2.11 (or on the left in Fig. 2.11(b)) consists of the bifurcation curves corresponding to the destruction of attractor S_{01} or to the formation of multiloop separatrices of saddle-focus O_2. The dashed line 3 within region D_1 corresponds to the first change in stability of cycle L_{01}.

For parameter values of region D_2, a stable oscillatory limit cycle L_{02} (Fig. 2.12(c)) exists in the phase space U. Line 4 corresponding to the bifurcation curve of cycle L_{02} is the boundary of region D_2. Similarly to cycle L_{01}, cycle L_{02} can undergo period doubling bifurcations that are completed with the generation of a chaotic oscillatory attractor S_{02} (Fig. 2.12(d)). The attractor S_{02} is realized at parameter values of region D_{H2}. The dashed line 5 within region D_2 corresponds to the first period doubling of cycle L_{02}. This line coincides with the dashed line 3 when $\gamma = 0$.

Region D_3 is a parametrical area, which allows for the existence of rotatory-type attractors in U phase space. This area lies

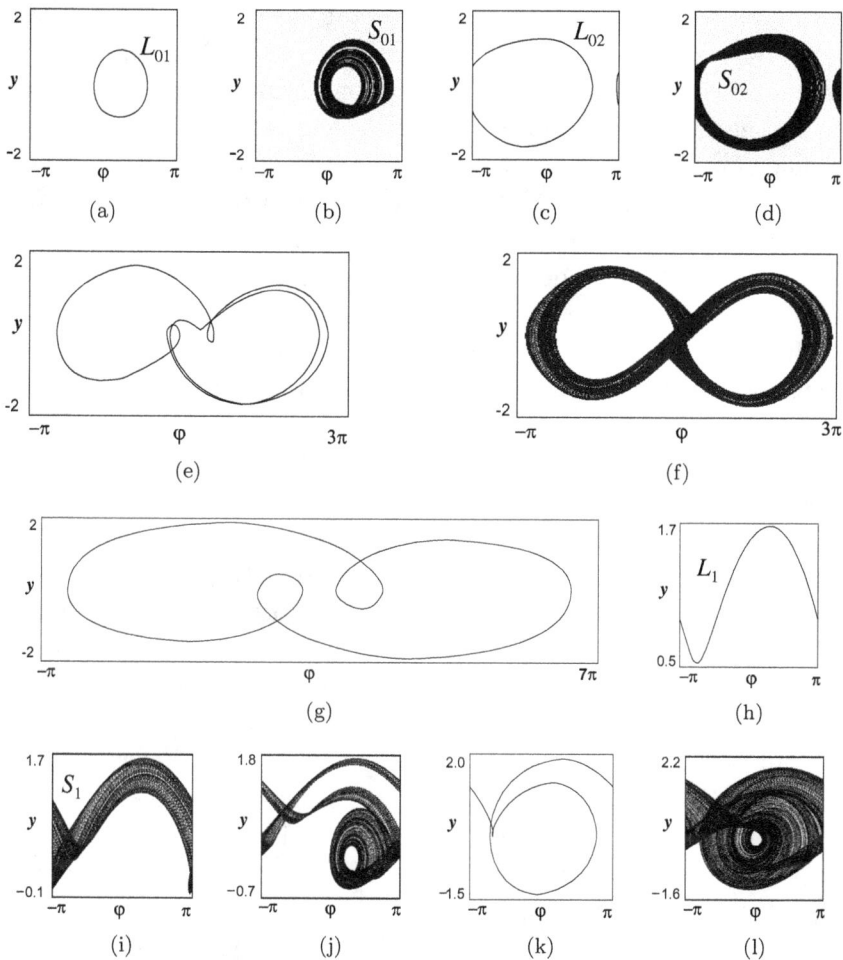

Fig. 2.12 The projections (φ, y) of the attractors of the model (2.3) for $\gamma \neq 0$.

between lines 6 and 7. Line 6 consists of two curves. One of the curves corresponds to the formation of a stable separatrix saddle loop (saddle-focus) O_2, which encircles phase cylinder U itself. The other corresponds to the saddle-node bifurcation of the rotatory cycle which encircles the phase cylinder itself. When entering region D_3 through line 6 in system phase space, we can observe the stable oscillatory limit cycle L_1 (Fig. 2.12(h)). Cycle L_1 can undergo a series of period doubling bifurcations and generate a chaotic rotatory attractor S_1 (Fig. 2.12(i)). Chaotic rotatory attractors of the system (2.3) exist at the parametrical values from region D_{H3}. Line 7 is composed of bifurcation curve portions, which correspond either to the generation of multiloop saddle-focus separatrices O_2, or to saddle-node bifurcations of high multiplicity (two, four or higher) stable rotatory cycles, or else to the chaotic rotatory attractor crisis. Hereinafter the terms "crisis" and "intrinsic bifurcation" of the attractor should be understood as in [51]. Dashed line 8 inside region D_3 corresponds to the first L_1 cycle period doubling.

In region G, the system (2.3) is characterized by complex dynamic behavior connected with various oscillatory–rotatory limit cycles and chaotic attractors. When moving inside these regions, oscillatory–rotatory attractors are undergoing constant evolution, passing from regular into chaotic and vice versa. During this process, the attractor structure changes, i.e., we can observe changes in the oscillatory–rotatory movement ratio. Figures 2.12(j–l) demonstrates the projections of regular and chaotic oscillatory–rotatory attractors of a different structure. Computational experiments show that purely oscillatory attractors, both regular and chaotic, can also be realized in region G. However, their regions are small and insignificant.

The complicated pattern of the systems dynamic behavior modifications (2.3) observed when the system moves inside region G is illustrated by Poincaré map single-parametric bifurcation diagrams $\{\gamma, \varphi\}$ represented in Fig. 2.13. The dark band above the diagrams characterizes the rotation along the coordinate φ. The absence of the dark band means that an oscillatory attractor is realized at these values of the parameter γ in the phase space U. From Fig. 2.13 it is clear that the dynamic behavior scenarios vary depending on the increase or decrease of the γ parameter. This fact testifies to the presence of multistable modes in the system (2.3). These modes allow for the co-existence of different types of attractors — regular as well as chaotic.

Oscillatory attractors with phase turns can be observed in model (2.3) at $\gamma \ll 1$. These movements are extremely diverse: depending on the filter parameters relationship, they can be regular or chaotic, symmetrical or asymmetrical, with one, two or more turns

(a)

(b)

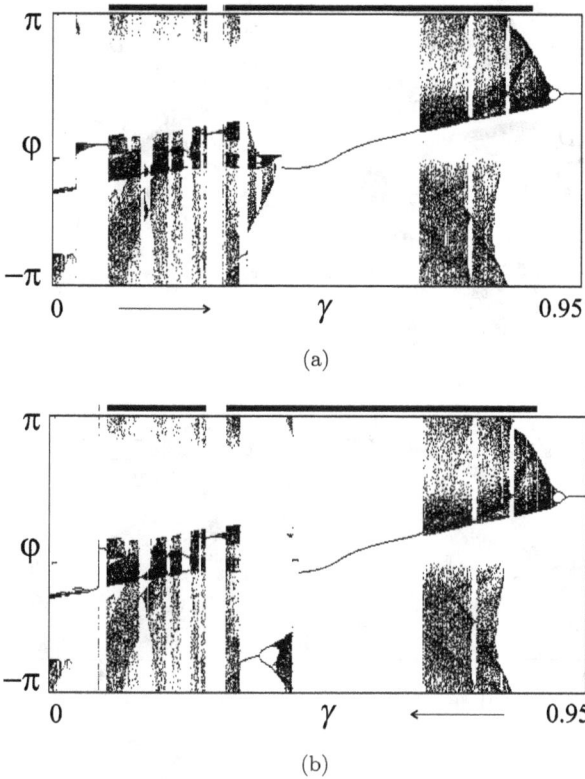

Fig. 2.13 Single-parametric Poincaré map of bifurcation diagrams of system (2.3), which, at $\varepsilon = 1$, $\mu = 4.1$, $n_1 = n_2 = 0$, is obtained by (a) increase and (b) decrease of the parameter γ.

(Figs. 2.12(e–g)). Regions of existence of oscillatory attractors with phase turns are insignificant, that is why they are not specified in Fig. 2.11.

Region D_Z in Fig. 2.11 provides pull-in to the synchronization mode. In this region, the equilibrium state is globally and asymptotically stable.

2.4.3.   *The phase system with a filter [2/2]*

Peculiarities of the dynamic behavior of the model (2.3) at $n_1 \neq 0$, $n_2 \neq 0$, $F(\varphi) = \sin \varphi$ are reflected in parametrical portraits (n_1, γ), (n_2, γ) and (μ, γ) represented in Fig. 2.14. At $n_1 \neq 0, n_2 \neq 0$, the

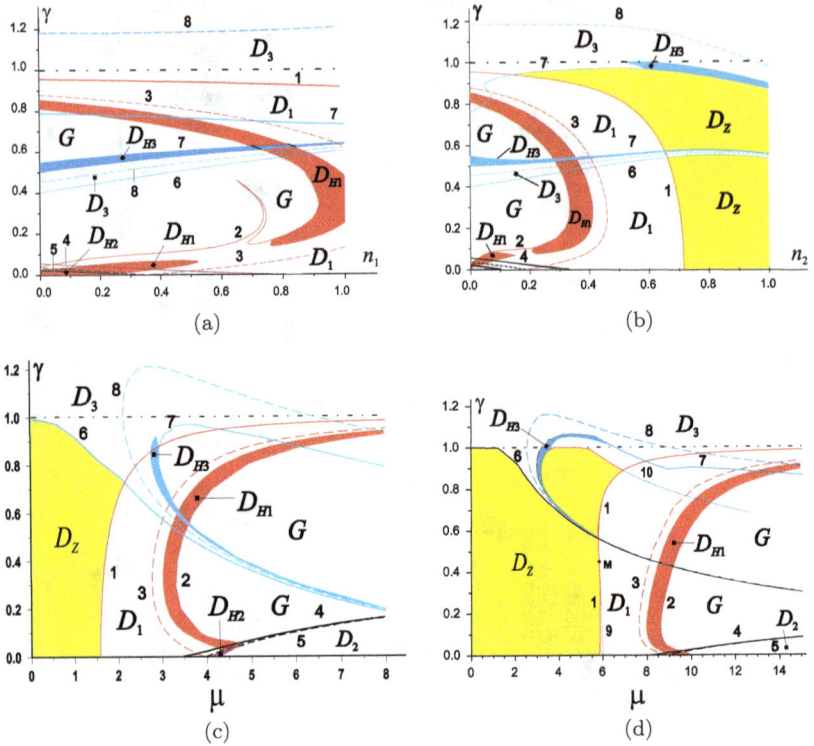

Fig. 2.14 The bifurcation diagram of the dynamic modes of model (2.3) if $F(\varphi) = \sin \varphi$ (a) $\varepsilon = 1$, $\mu = 3.5$, $n_2 = 0$; (b) $\varepsilon = 1$, $\mu = 3.5$, $n_1 = 0$; (c) $\varepsilon = 1$, $n_1 = n_2 = 0.2$; (d) $\varepsilon = 1$, $n_1 = n_2 = 0.5$.

structure of the parameter space partition previously established for the case $n_1 = 0, n_2 = 0$ is, in general, preserved. Reference points of the curves and regions in Fig. 2.14 coincide with reference points of curves and regions in Fig. 2.11. Let us consider the features, which appear at $n_1 \neq 0, n_2 \neq 0$.

The first feature is connected with the character of the synchronization mode existence and pull-in region. At $n_1 n_2 \neq 0$, the change in the stability of equilibrium state O_1, which corresponds to the synchronization mode, can be very hard. It reflects on the character of line 1, which restricts region D_Z. Now this line can consist of the sections of two bifurcation curves joined at the point M (Fig. 2.14(d)). The M point corresponds to the conversion of the first Lyapunov exponent $L = 0$ to zero. The line 1 portion located above the M point is determined by $\gamma = \gamma_s$ bifurcation curve, on which $L < 0$. For this reason, the entrance to D_1 region through this particular section is accompanied by a soft mode of generation of a stable oscillatory limit cycle L_{01} (Fig. 2.12). The line 1 portion located below the M point corresponds to saddle-node bifurcation, which results in the generation of stable L_{01} and saddle Γ_{01} oscillatory limit cycles. In this way, when the intersection of curve γ_s takes place below M point, the loss of synchronization mode stability and transition to quasisynchronous oscillations have a hard character and are accompanied with hysteretic phenomenon.

In the cases when $n_1 = 0$ or when $n_2 = 0$, the first Lyapunov exponent L is always negative. Consequently, line 1 consists of only $\gamma = \gamma_s$ bifurcation curve. The dangerous segment on the γ_s curve (a segment, where $L > 0$) appears at $n_1 n_2 \neq 0$ from the $\gamma = 0$ point and expands to increase the initial frequency detuning. Hereinafter, the terms "dangerous" and "safe" segments of stability change borders should be understood as in [52]. If at $\gamma = 0$, the first Lyapunov exponent $L < 0$, it is also negative at $\gamma \neq 0$. In this way, we can assess the presence of a dangerous segment on the stability change curve O_1 by the value of L exponent at $\gamma = 0$. If at $\gamma = 0$, $L < 0$, the whole γ_s curve is safe, whereas $L > 0$ indicates the presence of a dangerous segment on the γ_s border. To illustrate the possibilities in the occurrence of a dangerous border, Fig. 2.15 shows three- and two-dimensional projections of the $L(\gamma, \varepsilon, \mu, n_1, n_2) = 0$ surface for $\gamma = 0$. Interestingly, at $\gamma = 0$, $n_2 = 0.25$ the μ and ε parameters on the $L = 0$ surface have extrema, whose relation is a fixed value $\mu_{\min}/\varepsilon_{\max} = 4$, where $\varepsilon_{\max} = n_1^{-1}$.

The second feature refers to the D_Z synchronous mode pull-in region. Firstly, from the bifurcation diagrams presented in Fig. 2.14, it follows that the pull-in region can be considerably increased by matching n_1, n_2 parameters. This method permits to preserve the safety of synchronization mode stability change border. Secondly, at $n_1 > 0$ or $n_2 > 0$ the D_z region is composed of two sub-regions (Figs. 2.14(b) and 2.14(d)), divided by a narrow parameter band responsible for the existence of asynchronous movements in the U

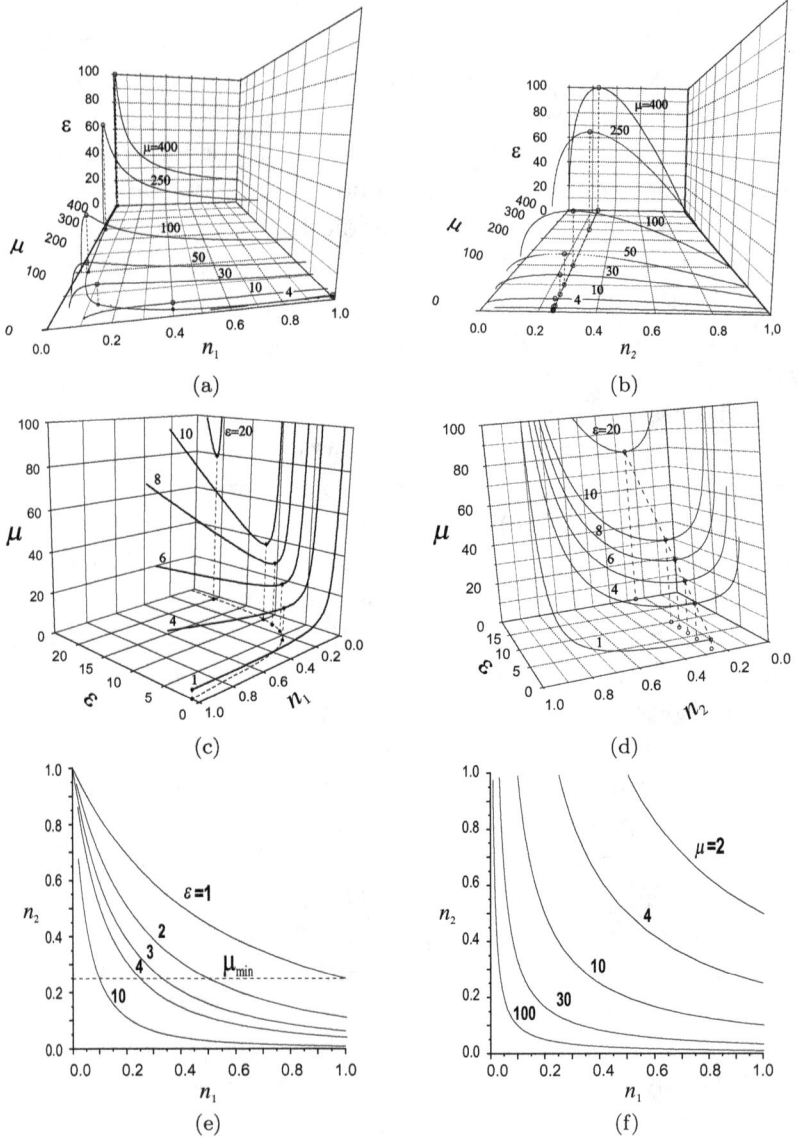

(a)

(b)

(c)

(d)

(e)

(f)

Fig. 2.15   Projections of the surface of the first Lyapunov exponent conversion to zero for the value $\gamma = 0$.

phase space. However, the dividing layer can be so narrow that D_Z region can practically exist as a single whole. The borders of the D_Z region run along lines 1, 6, 7 and line 10. The latter corresponds to the occurrence of oscillatory–rotatory type of attractors. At the values of the parameters from D_Z sub-region, which is entirely located within the area of high initial detuning, there are unstable regular movements in the U phase space. They can complicate and protract transition to the synchronization mode.

The introduction of n_1 and n_2 parameters permits to improve the situation with the generation of chaotic quasisynchronous oscillations. Specifically, it permits to confine the integration of D_{H1} sub-regions into a single region and, to some extent, expand chaotic quasisynchronous oscillation generation regions and their sections, in which the occurrence of this mode is guaranteed. At $n_1 \neq 0$, $n_2 \neq 0$, regular and chaotic quasisynchronization mode pull-in regions have the same features as the D_Z synchronization mode pull-in region, i.e., they can be composed of two sub-regions divided by a narrow asynchronous mode existence band.

In this way, we can affirm that bifurcation diagrams in Figs. 2.11 and 2.14 give a clear idea about the dynamics of a second-order filter PLL system and its evolution caused by the modification of the filter parameters. From the above-presented descriptions, we can see that the parameter space beyond the stability region has a complex structure. It is characterized by a large number of regions that correspond to different dynamic modes. These regions can overlap causing the multistable system behavior and lead to various hysteretic phenomena. Monostable behavior is most articulate in the D_Z pull-in region of synchronous mode and in the adjoining regions of quasisynchronous regular movements and regular beats. The region of chaotic quasisynchronous oscillations exist in a considerably small area. Moreover, this region, firstly, is composed of several sub-regions and, secondly, partly lies in the zone of multistable system behavior. All of the above factors impede the realization of the chaotic quasisynchronization mode. By changing the n_1 and n_2 parameters, it is possible to achieve a certain expansion of the parameter regions, which allow for the realization of synchronous and quasisynchronous modes.

2.4.4.   *The influence of the PD nonlinearity*

Now let us consider the influence of the PD nonlinear characteristic on the PLL system dynamics. Figure 2.16 shows the results of the examination carried out with regard to the (2.3) model dynamic modes for different PD characteristics in the case when $n_1 = 0$ and $n_2 = 0$. We have examined three particular types of characteristics symmetrical to the straight line, including $\varphi = \pi/2$: $F(\varphi) = \sin^3 \varphi$ (Fig. 2.16(a)), $F(\varphi) = \sin^{1/3} \varphi$ (Fig. 2.16(b)) a trapezoid-type characteristics defined by Eq. (2.4) (Figs. 2.16(b) and 2.16(d)), and asymmetric sawtooth-type PD defined by Eq. (2.5) (Fig. 2.16(e))

$$F(\varphi) = \begin{cases} \varphi/a, & -a < \varphi < a, \\ 1, & a \le \varphi \le \pi - a, \\ -1, & -\pi + a \le \varphi \le -a, \\ -(\varphi - \pi)/a, & \pi - a < \varphi < \pi, \\ -(\varphi + \pi)/a, & -\pi < \varphi < -\pi + a, \end{cases} \qquad (2.4)$$

$$F(\varphi) = \begin{cases} \varphi/b, & -b \le \varphi \le b, \\ (\varphi - \pi)/(b - \pi), & b < \varphi < \pi, \\ (\varphi + \pi)/(b - \pi), & -\pi < \varphi < -b. \end{cases} \qquad (2.5)$$

In Fig. 2.16, we have used the same denotations of the curves and regions as in Fig. 2.11. Comparing Figs. 2.11(a) and 2.16 leads to the conclusion that the modification of the nonlinear PD characteristic of the PLL system at $\varepsilon = 1$ does not result in the quality change of the parametrical portrait (μ, γ) — the disturbance of the synchronous mode stability, as well as the origination and chaotization of self-modulated oscillations happen in accordance with the scenarios, which were previously established for the sinusoidal PD characteristic. The major modifications are related to the shape and size of parameter regions corresponding of the existence of respective oscillations. Let us examine the results of the comparative analysis done for chaotic oscillation regions.

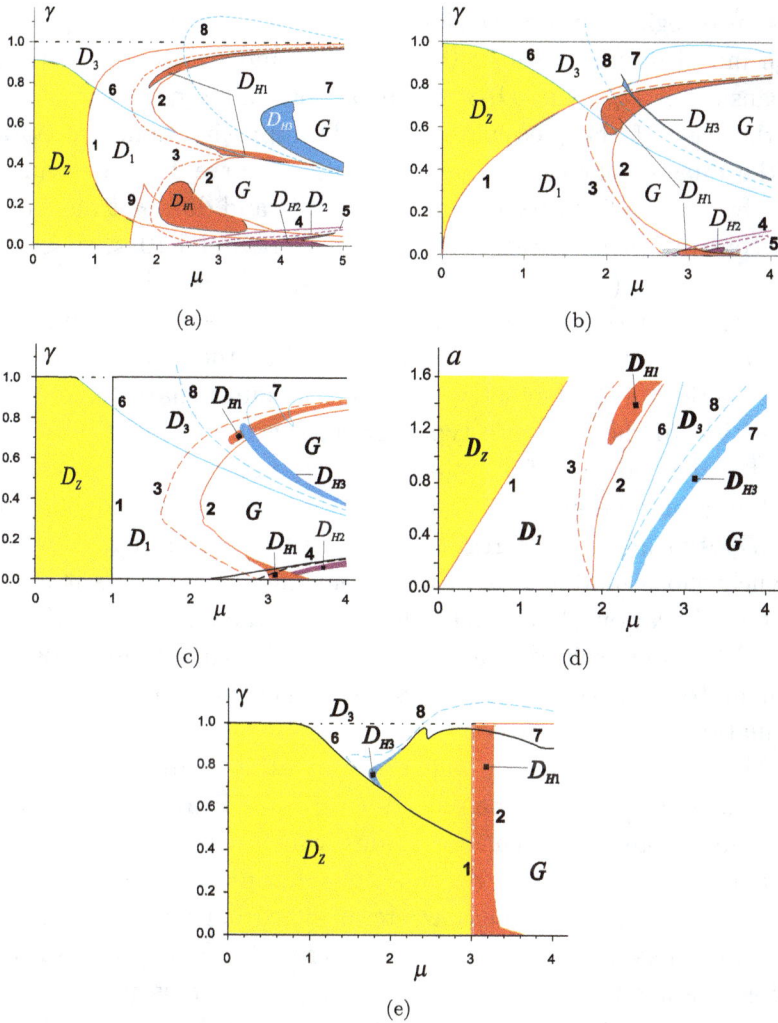

Fig. 2.16 Bifurcation diagrams of the dynamic modes for the second-order filter PLL (2.3) model of the [0/2] ($n_1 = n_2 = 0$) type at $\varepsilon = 1$, (a) $F(\varphi) = \sin^3 \varphi$; (b) $F(\varphi) = \sin^{1/3} \varphi$; for the trapezoid-type PD characteristic at (c) $a = 1$ and at (d) $\gamma = 0.5$; for the sawtooth-type PD characteristic at (e) $b = 3$.

The increase in the PD characteristic steepness in the zero position ($F(\varphi) = \sin \varphi^{1/3} \rightarrow F(\varphi) = \sin \varphi \rightarrow F(\varphi) = \sin^3 \varphi$ transition or a parameter reduction for the trapezoidal characteristic) leads to the reduction of the D_{H1} chaotic quasisynchronous mode

existence regions and expansion of the D_{H3} chaotic beat regions at
high initial detuning values $0.5 < \gamma < 1$, and, alternatively, to the
expansion of D_{H1} and D_{H2} regions at low values γ ($0 < \gamma < 0.5$).
In the case of the piecewise-linear PD approximation, the border
of the quasisynchronous mode generation (curve 1) is determined
only by the nonlinearity parameter (a or b) and does not depend on
the control circuit parameters. At the trapezoidal ($0 < a < \pi/2$)
and triangular ($a = \pi/2$) PD characteristics, when the characteristics are symmetrical with regard to $\varphi = \pi/2$, chaotically modulated
oscillations (CMO) are strongly influenced by the γ parameter. The
introduction of dissymmetry into the triangular characteristic, i.e.,
transition to the sawtooth-type PD characteristic, practically rules
out the dependence of the chaotic synchronization existence mode
on the γ parameter, and, at the same time, considerably expands
the D_Z synchronization region, reducing the region of regular quasisynchronous modes. In this way, from the above results it follows
that modification of the phase detector characteristic can influence
the PLL system self-oscillatory modes. At the same time it does not
contribute to the considerable expansion of chaotic quasisynchronous
oscillation generation regions.

The results obtained through the bifurcation analysis of the model
(2.3) demonstrate that the second-order filter PLL system abounds
in various regular and chaotic modes, which depend on the system
parameters, as well as on the initial conditions. We have discovered
new information about the character of synchronization mode pull-in
region borders. Namely, we have discovered that bifurcation surfaces
corresponding to chaotic attractor crisis can act as pull-in region
borders. It has been proved that by changing the form of the PD
characteristic and filter parameters, we can considerably improve the
synchronizing features of a PLL system. It has also been discovered
that the PLL system has a wide range of capabilities in the generation
of modulated oscillations of various degrees of complexity. It has been
established that existence regions of CMO in a second-order filter
PLL system are comparatively small. By changing the filter parameters, it is possible to expand the parameter regions, where the occurrence of the chaotic quasisynchronous oscillation mode is guaranteed.

However, a radical expansion of chaotic quasisynchronous mode existence regions cannot be achieved by a change in the filter parameters. Nor can it be achieved by changing the PD form. Additional study is required for the development of guidelines in using the discovered PLL system chaotic mode features.

2.5. Self-modulation Modes Generation Regions

Let us analyze the existence regions of asynchronous modes (regular or chaotic). In the U phase space of the model (2.3), these modes are associated with either a stable limit cycle or chaotic attractor movements: consequently, we have to deal with either regular or chaotic self-oscillations. We would like to remark that, in this case, oscillations at the PLL system output are oscillations with regular or chaotic angular modulation, which appear spontaneously due to nonlinearity of the PLL system and not caused by the external modulating signal. Hereinafter, in compliance, such modes will be called self-modulation modes.

All (2.3) modes can be interpreted from the point of view of the PLL system behavior. Synchronization mode, i.e., the mode, which provides the generation of oscillations synchronized by the reference signal at the PLL system output, corresponds to the stable state $O_1(\arcsin \gamma, 0, 0)$. The regular quasisynchronization mode, i.e., the mode, which provides generation of oscillations with periodic angular modulation, whose frequency coincides with the frequency of the reference signal at the PLL system output, corresponds to stable oscillatory limit cycle (which does not encircle the U phase cylinder itself and, which, as a rule, is localized around the O_1 equilibrium state). The chaotic quasisynchronization mode, i.e., the mode, which provides generation of oscillations with chaotic angular modulation, whose average frequency coincides with the frequency of the reference signal at the PLL system output, corresponds to the chaotic oscillatory attractor. Beat modes, which provide the generation of regular or chaotic angular modulation at the PLL system output, correspond to regular or chaotic rotatory (oscillatory–rotatory) attractors, respectively. However, the average frequency of these oscillations does not coincide with the frequency of the reference signal and can vary within a wide range of values.

Regarding a possible applied use of PLL chaotic modes, chaotic oscillations with the angular modulation, whose average frequency is stabilized with a reference signal, are of great interest. Hereinafter, we will call such oscillations generated at the PLL output — Chaotically Modulated Oscillations (CMO). Further we will focus our attention on this particular mode.

The change in the synchronous mode by self-modulation ones in the PLL system is caused either by the change in stability or by the disappearance of the synchronous mode. The change in the synchronous mode stability is accompanied by the generation of quasisynchronous oscillations; whereas the disappearance of the synchronous mode leads to the establishment of the beat mode. From the bifurcation analysis of the (2.3) model it follows that self-modulation modes can appear long before the disappearance of the synchronous regime and that at certain mode parameter values, there can be several modes capable of interfering with the stable operation of the PLL system. In such parameter regions, the PLL system is characterized by multistability. Bifurcation diagrams in Figs. 2.11–2.16 reflect the mechanisms, which determine the appearance of different dynamic modes and their evolution, and permit to evaluate regions of the established mode within the parameter space. However, they are unable to always identify dynamic modes change scenarios, which develop on the borders of the specified regions, since these borders run through multistable behavior areas. Single parametric Poincaré map bifurcation diagrams or double parametric dynamic mode maps, whose construction algorithm is based on the construction and analysis of Poincaré maps together with the calculation of the maximum Lyapunov exponent, also serve to assess the system behavior along the mentioned region borders. The dynamic mode map depends on the fixed mathematical model parameters, on the selection of a phase trajectory, which generates Poincaré map and on the algorithm of transition from one parametric space point to another. The selection of the map construction algorithm is determined by the tasks solved during the map construction. The flexibility of the algorithm used for the construction of double parametric dynamic mode maps enables researchers to model different system processes and obtain a sufficiently realistic picture of the examined object behavior.

Let us study the results obtained from the examination carried out with regard to two dynamic characteristics: CMO mode pull-in D_z^c region and CMO mode pull-out D_u^c region. One of the main tasks solved in the study of chaotic oscillation generators study is the examination of parameter regions, in which generators create chaotic oscillations. Parameter regions of mathematical generator models with chaotic attractors are useful for the theoretical evaluation of the above specified regions. When specifying parameter regions, their borders are often marked by extreme parametrical values, at which chaotic attractors originate, die, or, in certain cases, undergo inner bifurcation in mathematical model phase spaces. CMO generation regions specified in this way are shown in Figs. 2.11–2.17. However, these figures do not provide any information about the inner structure of the regions. Meanwhile, this information is extremely important for developing recommendations in the applied use of chaotic oscillations.

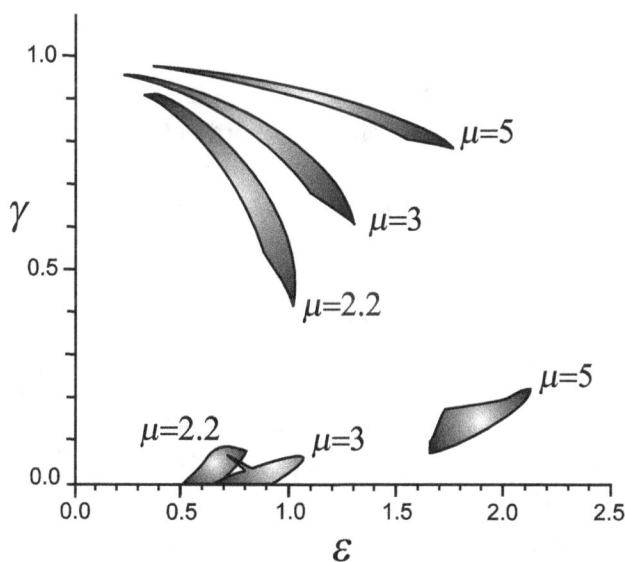

Fig. 2.17   Existence regions of chaotically modulated system oscillations (2.3).

We will evaluate D_u^c and D_z^c regions with algorithms similar to those normally used for the calculation of the band, in which PLL systems can be pull-in and pull-out in synchronous mode [41]. However, in this case, a quasisynchronous mode will be used instead of a synchronous one.

D_u^c **pull-out region of CMO mode.** Since quasisynchronous modes appear from a synchronous mode stability change, we can model the process of a PLL system exiting from the synchronous mode in the construction of a dynamic modes map characterizing the quasisynchronous modes pull-out region D_u^c. For the fixed values $\varepsilon = 1, n_1 = n_2 = 0$, let us assume that at $\gamma = 0.99$ the PLL system functions in the synchronization mode, i.e., the initial state of the (2.3) model lies within the range of a stable equilibrium state. We will gradually reduce further the γ parameter to 0 with the increment of $h_\gamma = 0.01$, using a trajectory, whose initial conditions belong to the last discovered attractor, in constructing the corresponding Poincaré map. This procedure will be repeated with the increment of $h_\mu = 0.01$ for all μ values ranging from 2 to 5. As a result, we will get a dynamic modes map (see Fig. 2.18), in which the type of the established dynamic mode will be highlighted using a specified color. This map gives a general idea of the dynamic modes distribution within the parameter space and reflects the results of the (2.3) model bifurcation analysis. It fully reflects the synchronization mode, regular and chaotic modes existence regions and represents asynchronous mode regions, that are determined by oscillatory and oscillatory–rotatory limit cycles and oscillatory–rotatory chaotic attractors.

Let us compare CMO generation regions represented in Fig. 2.18 with the regions represented in Fig. 2.11, obtained by bifurcation analysis. In Fig. 2.11 we have specified three CMO generation regions: the D_{H1} region located within the high initial detuning region and D_{H2} and D_{H3} regions located within the smaller initial detuning regions γ. D_{H2} and D_{H3} regions overlap in the $\mu \in (3.2; 3.7)$ interval and account for the existence of two different chaotic attractors with very similar features (Lyapunov exponent values, power range, autocorrelation functions). In Fig. 2.18, we can single out the D_{H1}^* region, which practically coincides with the D_{H1} region, and D_{H2}^*

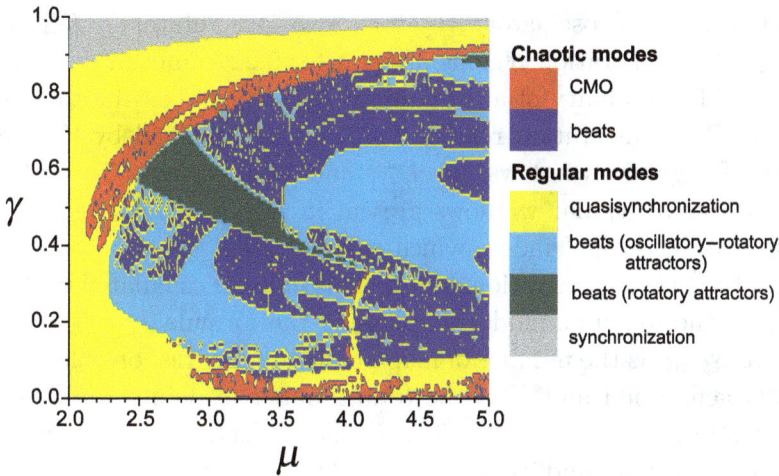

Fig. 2.18 Maps of dynamic modes of the PLL model (2.3), with filter [0/2] for $n_1 = n_2 = 0$, $\varepsilon = 1$.

region, which coincides with the shared section of D_{H2} and D_{H3} regions. Besides, the map shows previously unknown parameter values, which correspond to the CMO mode, and regular mode windows in D_{H1}^* and D_{H2}^* regions. In this way, the presented dynamic modes map reflects the established results of the chaotic mode research and supplements them with new data. It provides researchers with a simple method, that can be used to evaluate both the overall size of the region covering the existence regions of various dynamic modes and the size of each individually specified region. The overall size of the region, which covers dynamic modes of a certain type, is evaluated by calculating the proportion between the number of map points of a certain type and the sum total of points, which form the given map. A similar method is used to evaluate the size of individually specified regions — the number of points, which form the specified region, is compared to the total number of the map points. The overall size of dynamic mode regions in Fig. 2.18 is expressed in percentage terms: chaotic quasisynchronous mode regions account for 6.47% (oscillatory chaotic attractors); chaotic beats account for 27.08% (oscillatory–rotatory chaotic attractors); regular quasisynchronous oscillations account for 28%, whereas regular beats account for

34.54% (29.13% oscillatory–rotatory + 5.41% rotatory); D_{H1}^* and D_{H2}^* CMO generation regions account for 5.22% and 2.26% respectively. The summary of all the above regions exceeds the total size of the CMO generation region. This fact is explained by the presence of regular "windows" in D_{H1}^* and D_{H2}^* regions. To evaluate the degree of regular windows present in D_{H1}^* and D_{H2}^* regions we will introduce the I index, which characterizes "chaotic homogeneity" of the CMO generation region. This index is calculated with the help of the dynamic modes map, using the formula $I = N_{\text{CMO}}/N$, where N_{CMO} is the number of map points directly responsible for the CMO generation and N is the total number of points, which form the CMO generation region. The I index calculated in this way for the $D_{H1}^* = D_{H1}$ and $D_{H2}^* = D_{H2} \cup D_{H3}$ regions are, respectively, equal to $I_{H1}^* = 0.75$ and $I_{H2}^* = 0.65$.

The "chaotic homogeneity" index and the overall size of the CMO generation region comprise the characteristics of chaotic oscillations generation region in the parameter space (see Table 2.1) of the second-order filter PLL system regarding a number of PD characteristics, specifically, for $F(\varphi) = \sin^3(\varphi)$ (Fig. 2.16), $F(\varphi) = \sqrt[3]{\sin(\varphi)}$, (Fig. 2.16(b)), "trapezoidal" (Fig. 2.16(c)) and "sawtooth" (Fig. 2.16(e)) characteristics. The data in Table 2.1 have been obtained while calculating (μ, γ) plane dynamic mode maps at $\varepsilon = 1$, $0 \leq \gamma \leq 0.99$, $2 \leq \mu \leq 5$ for $F(\varphi) = \sqrt[3]{\sin(\varphi)}$, "trapezoidal" and "sawtooth" characteristics and $1.8 \leq \mu \leq 4.8$ for $F(\varphi) = \sin^3(\varphi)$, $\varepsilon_\lambda = 10^{-4}$. Since the change in the PD characteristic shape does not result in the structural change of the model (2.3) parameter space partition, we have preserved regional notations accepted for $F(\varphi) = \sin \varphi$ in Table 2.1. The only exception is the sawtooth characteristic represented by a single region D_{H1}, which is quite narrow as regards to μ, but lengthy with regard to γ and varies from 0 to 0.99. The data presented in Table 2.1 support the conclusion that by changing the phase detector characteristic we can influence the PLL system dynamic modes but cannot achieve a significant extension of the CMO generation regions.

D_z^c pull-in region of CMO mode. For assessing the D_z^c region, we will use the maps based on the following two algorithms.

Table 2.1   Characteristics of the PLL system CMO generation regions with a second-order filter.

Characteristics PD	Regions generation CMO	Sizes regions	Indexes chaotic homogeneity
$F(\varphi) = \sin \varphi$	D_{H1}	5.22%	0.75
	$D_{H2} \cup D_{H3}$	2.26%	0.65
$F(\varphi) = \sin^3 \varphi$	D_{H1}	1.7%	0.7
	$D_{H2} \cup D_{H3}$	8.76%	0.89
$F(\varphi) = \sin^{1/3} \varphi$	D_{H1}	3.98%	0.93
	$D_{H2} \cup D_{H3}$	0.67%	0.85
Trapezoidal type for $a = 1$	D_{H1}	2.34%	0.38
	$D_{H2} \cup D_{H3}$	1.82%	0.46
Sawtooth type for $b = 3$	D_{H1}	6.95%	0.84

The first algorithm is based on the definition of the model (2.3) parameters, where we observe the disappearance of asynchronous movements impeding the PLL system entry into the quasisynchronous mode. When this algorithm is realized, the initial values of the (2.3) model parameters enter the region of the beat mode. Then they leave this region to enter that of quasisynchronous modes. In this case, the phase space point belonging to the latest discovered attractor is used in the construction of Poincaré maps.

The second algorithm is based on the facts that at $t = 0$, the phase variable y takes on the value of the initial frequency detuning $y_0 = \gamma$ and that the phase variable φ is not controlled by $\varphi_0 \in [-\pi, \pi]$, whereas $z_0 = 0$. In this algorithm, the identification of the dynamic mode for each parameter space point is carried out not on the basis of a single-phase trajectory, but on a whole series of phase trajectories, whose initial conditions are evenly distributed along the $l_0 = \{-\pi < \varphi < \pi, y = \gamma, z = 0\}$ interval. If some of the phase trajectories of the said series are drawn to various attractors, it is assumed that the analyzed spatial point is the point of the PLL system multistable behavior.

In Fig. 2.19 we can see dynamic mode maps constructed for analyzing the D_z^c pull-in region with the first and second algorithms, respectively. While constructing the map, the γ parameter

(a)

(b)

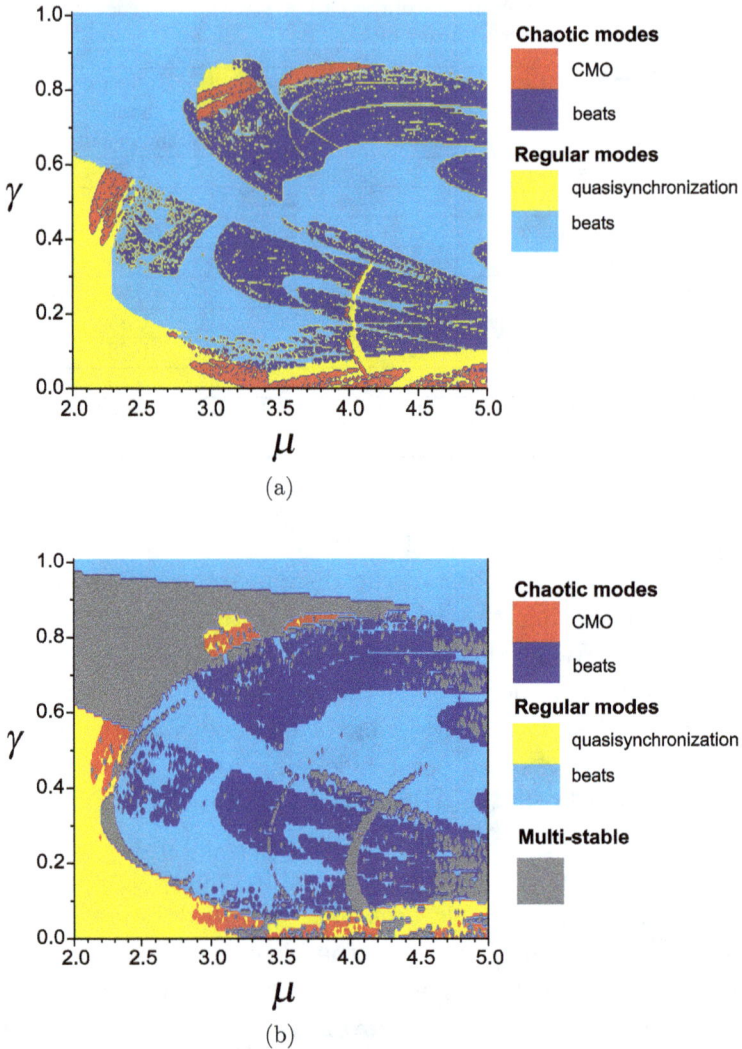

Fig. 2.19 The map of dynamical regimes of the model (2.3) for the assessment of CMO mode pull-in region if $n_1 = n_2 = 0$, $\varepsilon = 1$: (a) *the first algorithm*, (b) *the second algorithm*.

was changed from 1.01 to 0 by increasing $h_\gamma = 0.01$, while the μ parameter was changed from 2 to 5 with $h_\mu = 0.01$ increasing similarly. We can see that both algorithms used in calculating the D_z^c region give similar results. This conclusion is further supported by the graphic representation of the maps and by the results of the CMO mode pull-in region calculations, which are equal to 3.80% (a) and 3.59% (b), respectively. The peculiarity of the calculated CMO generation pull-in regions is the fact that as comparing the CMO holding mode region, they are even more fractured. In these regions, guaranteed establishment of the CMO mode is possible for a limited number of parameter values from the D^*_{H1} region. The CMO mode can be established in three sub-regions, $D^1_{z1}, D^2_{z1}, D^3_{z1}$ — of the D^*_{H1} region. They are very small and amount to 1.07%, 0.79% and 0.48%, respectively. The D_{z2} CMO pull-in regions within the incremental frequency detuning range is equal to 1.46%, which is less than the overall size of the D^*_{H2} region. In this way, we can state that the CMO mode pull-in regions of the second-order filter PLL system with a sine-wave PD characteristic is almost twice as small as CMO mode holding regions. This conclusion holds for other PD characteristics. Table 2.2 we can see the results of the CMO mode pull-in regions for the second-order filter PLL on the (μ, γ) plane at $\varepsilon = 1$, $0 \leq \gamma \leq 1.01$, $2 \leq \mu \leq 5$ for $F(\varphi) = \sqrt[3]{\sin(\varphi)}$, "trapezoidal" and "sawtooth" characteristics and $1.8 \leq \mu \leq 4.8$ for $F(\varphi) = \sin^3(\varphi)$. The calculation is based on the second algorithm. Table 2.2 shows only those pull-in areas, whose size exceeds 1%.

In Fig. 2.20 we can see a maximum Lyapunov exponent map, which corresponds to Fig. 2.19(a). Comparing the dynamic modes map with the maximum Lyapunov exponent map, it follows that asynchronous movements of the oscillatory–rotatory type have the

Table 2.2 Characteristics of the generation mode pull-in regions CMO of the second-order filter PLL system.

Characteristics $F(\varphi)$	Pull-in regions into the CMO mode	Sizes regions
$F(\varphi) = \sin \varphi$	D_{z2}	1.46%
$F(\varphi) = \sin^3 \varphi$	D_{z2}	3.34%
$F(\varphi) = \sin^{1/3} \varphi$	D^1_{z1}	1.64%
Trapezoidal for $a = 1$	D_{z2}	1.26%
Sawtooth for $b = 3$	D_{z1}	5.27%

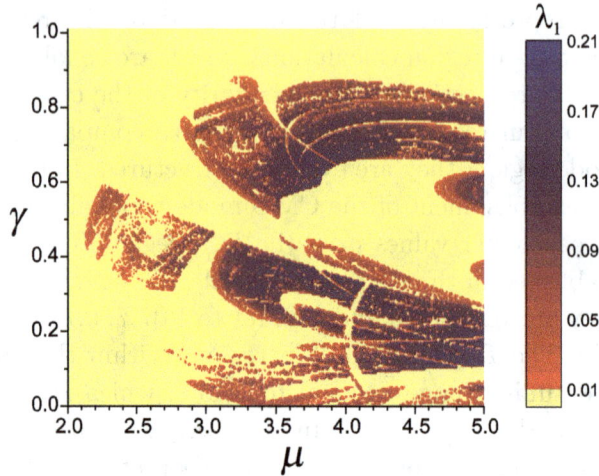

Fig. 2.20 The maximum Lyapunov exponent map, which corresponds to Fig. 2.19(a).

highest chaos degree in the model (2.3), as λ_1 takes on its maximum values on these attractors. Although maximum Lyapunov exponent maps clearly identify regions with chaotic behavior, they do not reflect the peculiarities of self-modulation modes. To be more specific, they do not allow for further classification of chaotic modes into quasisynchronous and asynchronous (beat modes).

2.6. Characteristics of Self-Modulation Modes

Self-modulation oscillations at the PLL system controlled generator output are caused by modulating oscillations in the control circuit. In the U phase space, they are represented by stable limit cycles and chaotic attractors. Consequently, chaotic attractor characteristics can be regarded as those of modulating oscillations. One of them is Lyapunov exponent (Fig. 2.20), which characterizes the degree of chaotic behavior demonstrated by modulating oscillations. Averaged values and ranges of phase coordinates on the (2.3) model stationary trajectories can be regarded as additional characteristics. Averaged values of oscillatory phase modulates can be obtained as average values of the φ and y coordinates, calculated for the limit cycle movement over a given period or for chaotic attractor movement over a considerably long period (typical of a given attractor). Maximum deviations of the φ and y coordinates from the averaged values observed relating to the limit cycle or chaotic attractor movement characterize the variation ranges of the said coordinates — $\Delta\varphi$ and Δy, respectively. Since modulation laws are determined by the model (2.3) solutions and since this model allows for multistable behavior, these laws may not necessarily be single valued. In other words, they may depend on the initial conditions, as well as on the model parameters. Figure 2.21, which shows the characteristics of modulating oscillations calculated for the dynamic modes represented in Fig. 2.18, illustrates this idea. The average deviation values and the $\Delta\varphi$ and Δy modulating oscillations amplitude are highlighted in

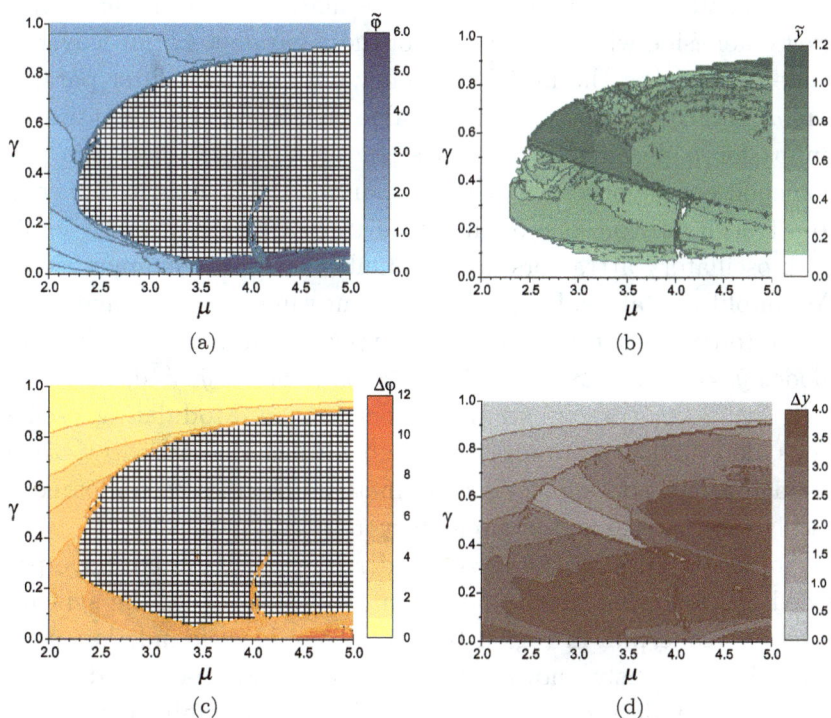

Fig. 2.21 The characteristics of modulating oscillations calculated for the dynamic modes represented in Fig. 2.18.

different colors. It should be noted that in this mode the φ coordinate is continuously growing. Consequently, there are no stationary average $\tilde{\varphi}$ values, or stationary $\Delta\varphi$ deviations, that is why in Figs. 2.21(a) and 2.21(c) the parameter space region corresponding to the beat modes is cross-hatched. Let us consider the existence regions of synchronous, quasisynchronous, phase turn quasi-synchronous and beat modes. However, average values and the φ and y variables amplitude are not sufficient to identify whether the output oscillations are chaotic or not.

There are no modulating oscillations in the synchronization mode, that is why $\Delta\varphi$ and Δy are equal to zero and the average values are consequently equal to $\tilde{y} = 0$ and $\tilde{\varphi} = \arcsin\gamma$. In the quasisynchronous mode, $\tilde{y} = 0$, $0 < \Delta\varphi < 2\pi$ and $\Delta y > 0$. We would like to draw the readers' attention to the following characteristics demonstrated by quasisynchronous oscillations: the $\Delta\varphi$ amplitude grows alongside with the decrease of the γ parameter; the $\tilde{\varphi}$ average value depends on the initial frequency detuning and filter parameters. Significantly, the $\tilde{\varphi}$ average value can change in discontinuous jumps for μ parameter variations. In the analyzed case this phenomenon can be observed at small γ and $\mu \approx 3.5$. This is connected to a transition from one oscillatory attractor to the other. Notably, both oscillatory attractors have a number of similar characteristics: $\Delta\varphi$ amplitude values, Lyapunov exponent values, power range values and auto-correlation functions. In the phase crank quasisynchronous mode, $\tilde{y} = 0$, $\Delta\varphi > 2\pi$ and $\Delta y > 0$, whereas $\tilde{y} \neq 0$, $|\tilde{\varphi}| > 2\pi$ and $\Delta\varphi > 2\pi$ correspond to the beat mode. Modulating oscillations, determined by rotatory attractors are characterized by bigger average values and smaller y coordinate amplitudes as compared to oscillations determined by oscillatory–rotatory attractors.

So, the studies on the second-order filter PLL system have provided the following new information on the nature of the synchronization mode pull-in region borders. They have found that apart from the previously known bifurcation separatrix loops and double limit cycles, pull-in region borders can be also determined by chaotic attractor crises. They show that the PLL system synchronization qualities can be significantly improved by changing the phase detector characteristic shape and varying the filter parameters.

Ample opportunities for generating modulating oscillations of different complexity have been discovered. PLL system self-modulation modes were classified. Among the chaotic oscillations we have singled out those of most practical interest — chaotic angular modulation oscillations with the averaged frequency stabilized against the reference frequency (CMO).

Algorithms for the identification of parameters, which characterize regions with different dynamic behavior, were suggested. We have used the suggested parameters to construct maps of the three-dimensional PLL model dynamic modes, which can be used to evaluate the CMO mode pull-out and pull-in regions for different PD characteristics. We have compared the results of the dynamic mode maps and those of the PLL system's mathematical model bifurcation analysis. It was discovered that, firstly, the maps match the results of the PLL system's mathematical model bifurcation analysis and, secondly, they help evaluate the size and the inner structure of the CMO generation regions. The results obtained in this section confirm the previously made conclusions about the diversity of the three-dimensional PLL model dynamic modes, a high degree of this model multistability and the subsequent presence of hysteretic phenomena.

Characteristics of modulating oscillations for the constructed dynamic mode maps were calculated. It was established that the analysis of Lyapunov exponent maps is not sufficient for the evaluation of the PLL system chaotic features, as the maps lack information on the structure of chaotic attractors and, consequently, the type of the dynamic mode.

We have carried out the analysis of the pull-out and pull-in CMO mode regions for various characteristics of the PD and filter parameters. It has been established that the CMO existence regions in the second-order PLL system are small and their chaotic homogeneity degree is comparatively low. In the next chapters, we will consider the possibilities for the expansion of the CMO existence regions by transiting from isolated PLL to an ensemble of coupled PLLs.

Chapter 3

Cascade Coupling of Two Phase Systems

3.1.   Ensemble Mathematical Models

To connect the phase-locked loop (PLL) system chains into an ensemble, the outlet of the previous PLL system will become an input of the following PLL system [4, 31]. For the chain connection of the PLL systems, using a single external signal is typical. The suitability of a cascade connection of systems is discussed in [30]. By such connection, PLL can improve filtration of interfering signals. Actually, by considering that each PLL system can be viewed as a filter for a signal coming into its input, then at the series connection the filtering property of the filter chain is more than the property of one element of the chain. The presence of the local control circuit allows us to organize additional coupling between the elements of the circuit not only to change dynamic properties of the joined systems but also to control these properties through changing coupling parameters.

Figure 3.1 shows a structural diagram of the ensemble of two cascade-connected PLL systems. Here, an external signal is a reference signal $S_0(t)$, which comes into the input of PLL_1, and signal $S_1(t)$ from the outlet of PLL_1 is an input signal for PLL_2. Partial systems PLL_1 and PLL_2 can interchange information about error signals e_1 and e_2, appearing in them through the circuits of mutual control CC_{12} and CC_{21}. Control circuit CC_{12} carries out an additional coupling "forward" passing information from the outlet of the phase discriminator FD_1 to the local control circuit PLL_2. Control circuit

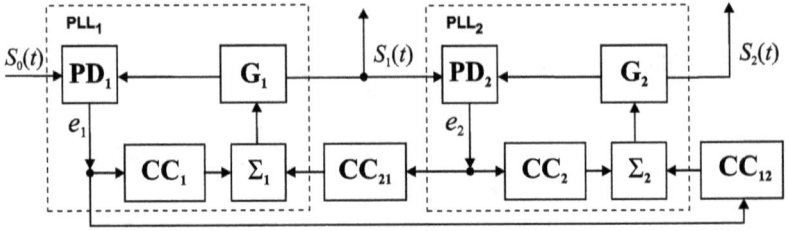

Fig. 3.1   Structural diagram of the cascade connection of two PLLs.

CC_{21} carries out an additional coupling "backward" passing information from the outlet of the phase discriminator FD_2 to the local control path PLL_1. Signals in the control circuits CC_{12} and CC_{21} can be changed — be amplified or decay — and they go through additional filters.

The equations describing dynamic processes in the examined cascade system are formed of general equations of interconnected synchronized systems. General equations are usually written in an operator form and for a cascade connection of two PLLs, they have the following forms:

$$\frac{p\psi_1}{\Omega_1} = \gamma_1 - K_1(p)F_1(\psi_1) - \kappa K_{21}(p)F_2(\psi_2),$$

$$\frac{p\psi_2}{\Omega_1} = \gamma_2 - bK_2(p)F_2(\psi_2) - \delta K_{12}(p)F_1(\psi_1) - \frac{p\psi_1}{\Omega_1} \qquad (3.1)$$

or

$$\frac{p\varphi_1}{\Omega_1} = \gamma_1 - K_1(p)F_1(\varphi_1) - \kappa K_{21}(p)F_2(\varphi_2 - \varphi_1),$$

$$\frac{p\varphi_2}{\Omega_1} = \gamma_2 - bK_2(p)F_2(\varphi_2 - \varphi_1) - \delta K_{12}(p)F_1(\varphi_1). \qquad (3.2)$$

Here, $p \equiv d/dt$, $\psi_1 = \varphi_1 = \theta_1 - \theta_0$ is the current phase error between a signal of the first generator and a reference signal, $\psi_2 = \theta_2 - \theta_1$ is the current phase error between the signals of the first and the second generators, $\varphi_2 = \psi_1 + \psi_2 = \theta_2 - \theta_0$ is the current phase error between a signal of the second generator and a reference signal, γ_1 and γ_2 are initial frequency detunings of the second and the first generators for a reference signal, $b = \Omega_2/\Omega_1$, where Ω_1 and Ω_2 are pull-out bands of the partial systems PLL_1 and PLL_2, $F_1(\psi_1)$ and $F_2(\psi_2)$ are standard characteristics of phase discriminators PLL_1 and PLL_2, $K_1(p)$, $K_2(p)$ and $K_{12}(p)$, $K_{21}(p)$ are transfer functions of low-frequency filters in local control circuits PLL_1 and PLL_2 and in circuits of additional couplings "forward" and "backward", δ and κ are transformation parameters of error signals between generators in forward and backward directions, respectively. Substituting actual terms of filter gains, characteristics of phase discriminations into Eq. (3.1) or (3.2), it is possible to get various mathematical models of two-circuit interrelated PLLs. We restrict ourselves to viewing those models which are obtained from Eq. (3.2) with the following conditions: $F_1(\cdot) = F_2(\cdot) = \sin(\cdot)$, $\Omega_1 = \Omega_2 = \Omega$, in control circuits there exist first- or second-order filters, where filter gains in the additional control circuits coincide with the terms for the filters in local control circuits. By considering the performed conditions from Eq. (3.2), the following mathematical models are obtained:

- for the second-order filters with transmission factors $K_1(p) = K_{12}(p) = (1 + a_1 p + b_1 p^2)^{-1}$ and $K_2(p) = K_{21}(p) = (1 + a_2 p + b_2 p^2)^{-1}$, the system of ordinary differential equations

$$\frac{d\varphi_1}{d\tau} = y_1, \quad \frac{dy_1}{d\tau} = z_1,$$

$$\mu_1 \frac{dz_1}{d\tau} = \gamma_1 - y_1 - \varepsilon_1 z_1 - \sin\varphi_1 - \kappa \sin(\varphi_2 - \varphi_1),$$

$$\frac{d\varphi_2}{d\tau} = y_2, \quad \frac{dy_2}{d\tau} = z_2,$$

$$\mu_2 \frac{dz_2}{d\tau} = \gamma_2 - y_2 - \varepsilon_2 z_2 - \sin(\varphi_2 - \varphi_1) - \delta \sin\varphi_1, \quad (3.3)$$

  is described in a six-dimensional cylinder phase region $U = \{\varphi_1(\text{mod}\,2\pi), \ y_1, z_1, \varphi_2(\text{mod}\,2\pi), y_2, z_2\}$. Here, $\tau = \Omega t$ is the dimensionless time, $\varepsilon_j = \Omega a_j$, $\mu_j = \Omega^2 b_j$ are dimensionless filters parameters, $j = 1, 2$.

- for the first-order filters in the case $K_1(p) = K_{21}(p) = (1 + m_1 T_1 p)(1 + T_1 p)^{-1}$, $K_2(p) = K_{12}(p) = (1 + m_2 T_2 p)(1 + T_2 p)^{-1}$, the system of equations

$$\frac{d\varphi_1}{d\tau} = u_1 - m_1 \left[\sin \varphi_1 + \kappa \sin(\varphi_2 - \varphi_1)\right],$$

$$\varepsilon_1 \frac{du_1}{d\tau} = \gamma_1 - u_1 - (1 - m_1) \left[\sin \varphi_1 + \kappa \sin(\varphi_2 - \varphi_1)\right]$$

$$\equiv P(\varphi_1, u_1, \varphi_2, u_2),$$

$$\frac{d\varphi_2}{d\tau} = u_2 - m_2 \left[\sin(\varphi_2 - \varphi_1) + \delta \sin \varphi_1\right],$$

$$\varepsilon_2 \frac{du_2}{d\tau} = \gamma_2 - u_2 - (1 - m_2) \left[\sin(\varphi_2 - \varphi_1) + \delta \sin \varphi_1\right]$$

$$\equiv Q(\varphi_1, u_1, \varphi_2, u_2), \tag{3.4}$$

is described in a four-dimensional cylinder phase region $V = \{\varphi_1 \pmod{2\pi}, u_1, \varphi_2 \pmod{2\pi}, u_2\}$. Here, $\tau = \Omega t$ is the dimensionless time, $\varepsilon_j = \Omega T_j$, $0 \le m_j < 1$, are dimensionless parameters of filters, $j = 1, 2$.

At $\varepsilon_1 \ll 1, \varepsilon_2 \ll 1$, the system (3.4) is a dynamic system with small parameters with derivatives $du_1/d\tau, du_2/d\tau$. General motion in the phase space V is divided into "rapid" and "slow" motions. Since

$$\frac{\partial P}{\partial u_1} + \frac{\partial Q}{\partial u_2} = -2 < 0, \qquad \frac{\partial P}{\partial u_1}\frac{\partial Q}{\partial u_2} - \frac{\partial P}{\partial u_2}\frac{\partial Q}{\partial u_1} = 1 > 0,$$

the toroidal surface $W : \{P(\varphi_1, u_1, \varphi_2, u_2) = 0, Q(\varphi_1, u_1, \varphi_2, u_2) = 0)\}$ of the slow motions is stable with respect to rapid motions. The equations of slow motions on the surface W are given by

$$\frac{d\varphi_1}{d\tau} = \gamma_1 - \sin \varphi_1 - \kappa \sin(\varphi_2 - \varphi_1),$$

$$\frac{d\varphi_2}{d\tau} = \gamma_2 - \sin(\varphi_2 - \varphi_1) - \delta \sin \varphi_1. \tag{3.5}$$

The periodicity of the variables φ_1 and φ_2 on the right-hand side of the system (3.5) with the period 2π indicates that it is a nonlinear dynamic system on torus $T = \{\varphi_1(\mathrm{mod}\,2\pi), \varphi_2(\mathrm{mod}\,2\pi)\}$. It defines dynamic processes of the cascade system with low-inertia control circuits.

Models (3.3)–(3.5) are written based on further analysis of the dynamic properties of the ensemble of two cascade-connected PLLs.

3.2. Classification of Dynamic Modes

Analysis of dynamic properties of the examined systems according to their mathematical models is based on the one-to-one correspondence between dynamic modes of the systems and attractors of mathematical models defined in the corresponding phase spaces. Phase spaces of the models of PLL have a distinctive feature: they contain the angular coordinate φ. It reflects on the behavior pattern of phase trajectories and on dynamic modes of PLL systems, respectively.

Models of a single PLL are defined in cylindrical phase spaces with one angular coordinate φ. Attractors which exist in these spaces are divided into two classes: attractors which encircle the phase cylinder and attractors that do not encircle the phase cylinder. Attractors which do not encircle the phase cylinder (restricted at the angular coordinate φ) are called oscillating attractors. They correspond to the quasisynchronous modes which at the PLL outlet are modulated by angle oscillations and medium frequency stabilized by the reference signal (with the exception of stable equilibriums corresponding to the pull-in mode of generator by a reference signal.)

Attractors which encircle the phase cylinder (non-limited at φ) are called rotatory attractors and correspond to the beating modes in which the phase difference of tuning and reference signals is constantly growing and the average difference of frequencies is equal to some constant value.

Models which describe collective dynamics of ensembles from n PLL systems are defined in the cylinder phase spaces containing n angular coordinates. Thereby, attractors of such models can be conditionally divided into 2^n classes which define physical modes of the ensemble's collective behavior and also the chaotic modes among them [53]. Collective dynamics of the ensembles PLL is defined by attractors in global phase spaces of the corresponding mathematical modes. Dynamics of the separate PLL which is a part of the ensemble is defined by the projections of the attractors on the corresponding partial subspaces.

For the examined ensemble which consists of two PLLs with second-order filter, the dynamics of PLL_1 depends on a kind of projection of model attractor (3.3) on the partial subspace $U_1 = \{\varphi_1(\mod 2\pi), y_1, z_1\}$, PLL_2 depends on a kind of projection of attractor on the partial subspace $U_2 = \{\varphi_2(\mod 2\pi), y_2, z_2\}$. The number of dimensions of a partial subspace is defined by the number of phase variables with the help of which the state of a separate PLL system in the ensemble is described. When the state of j generator is characterized by three phase variables φ_j, y_j, z_j in the model (3.3), then the number of dimensions of the partial subspaces is given by $\dim(U_j) = 3$. In the model (3.4), the state of j generator is characterized by two phase variables φ_j and u_j, and in the model (3.5) it is characterized by one phase variable φ_j. That is why numbers of dimensions of the partial subspaces for these models are, respectively, equal to $\dim(V_j) = 2$ and $\dim(T_j) = 1$, where $j = 1, 2$.

If an attractor projection on a corresponding partial subspace is limited by φ_j, the jth PLL system functions either in a synchronous or in a quasisynchronous mode, otherwise, the jth PLL system is in a mode of beatings. For PLLs which are in synchronous modes projection of attractors on the corresponding partial subspaces is related, to a point. Collective behavior of an ensemble which consists of n PLLs is labelled by a rotatory index (a quasisynchronous index) $[I_1, I_2, I_3, \ldots, I_n]$, where $I_j = 1$, if on the coordinate φ_j there is a rotation, and $I_j = 0$ if not. Non-existence of rotation by the coordinate φ_j means that the jth generators work in a quasisynchronous mode when the rest of the generators are in a mode of beatings. If an attractor is chaotic then at the outlet of the jth generator, there are either CMO (at $I_j = 0$) or chaotic beatings (at $I_j = 1$).

Figure 3.2 shows, as an example, two-dimensional projections of some attractors of the model (3.3) reproducing different modes of collective behavior of the examined ensemble. The attractors in

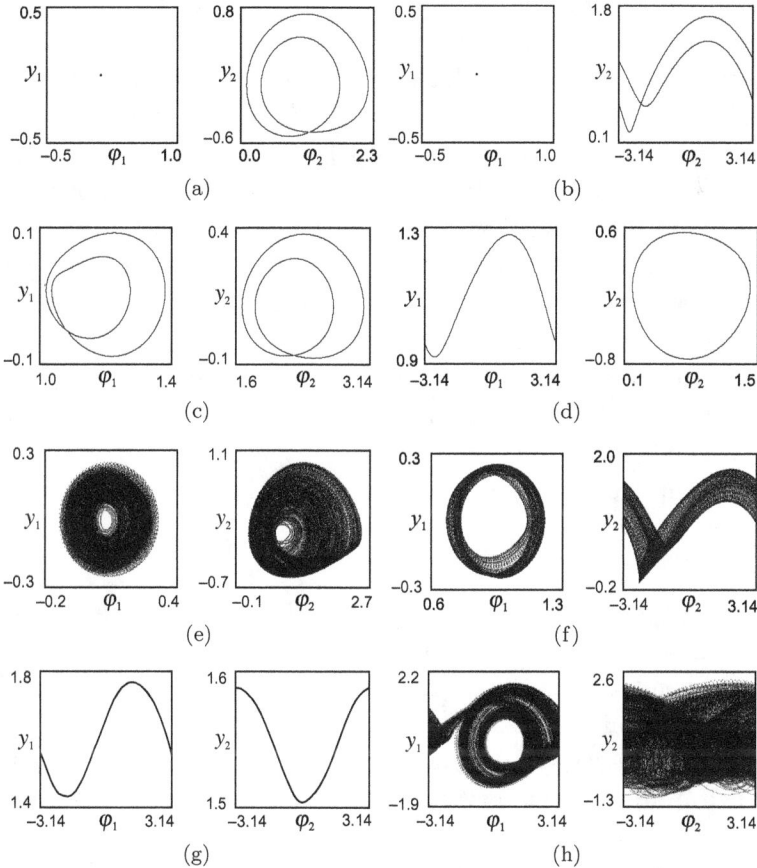

Fig. 3.2 Examples of projections of model attractors (3.3).

Figs. 3.2(a), 3.2(c) and 3.2(e) have a rotation index $[0, 0]$. In this connection, Fig. 3.2(a) corresponds to the synchronous mode PLL_1 and the regular quasisynchronous mode PLL_2; Fig. 3.2(c) to the regular quasisynchronous modes PLL_1 and PLL_2; and Fig. 3.2(e) to CMO mode at the outlet of PLL_1 and PLL_2. The attractors in Figs. 3.2(b) and 3.2(f) have a rotation index $[0, 1]$. Here, Fig. 3.2(b) corresponds to the synchronous mode PLL_1 and the mode of regular beatings PLL_2, Fig. 3.2(f) represents the CMO mode in the PLL_1 system and the mode of chaotic beatings in PLL_2. The attractor with index $[1, 0]$ in Fig. 3.2(d) shows the mode of regular beatings in the PLL_1

system and the regular quasisynchronous mode PLL_2. After all, the attractors in Figs. 3.2(g) and 3.2(h) have rotation index [1, 1] and correspond to regular and chaotic modes of beatings in PLL_1 as well as in PLL_2. From the presented figures, it can be seen that the system of two cascade-connected PLLs has a great range of modes of collective behavior. In this connection, it is not always possible to distinguish these modes by a kind of projections of attractors on the partial subspaces. For a clear determination of a dynamic mode type, it is necessary to involve additional computational procedures, for example, calculation of Lyapunov exponents. Determination of rotation index combined with determination of the maximum Lyapunov exponent allows us to identically decide the mode of functioning of the ensemble generators.

As mentioned above, dynamic modes of the ensemble can be classified as follows:

- *mode of global synchronization* — all controlled generators are synchronized relatively to a reference signal. Stable equilibriums correspond to this mode in the phase spaces of corresponding mathematical models.
- *mode of partial synchronization* — only separate generators are synchronized. Oscillatory–rotatory attractors, which have even one projection on the partial subspaces are generated into a point, correspond to these modes.
- *mode of global quasisynchronization* — at the outlet of all generators of the ensemble, they are modulated by an angle oscillation with average frequency which is stabilized by the reference frequency. Oscillatory attractors correspond to this mode.
- *mode of partial quasisynchronization* — among the generators of the ensemble even if one is in the quasisynchronous mode with respect to a reference signal, then the other function is in the mode of beatings. In the phase spaces of dynamic systems, oscillatory–rotatory attractors respond to these modes.
- *global mode of beatings* — this refers to all generators of the ensemble function in the mode of beatings. Rotatory attractors respond to these modes in the phase spaces of the dynamic systems.

Considering that attractors in the phase spaces of the mathematical models of the ensemble can be regular and chaotic, then the corresponding modes can also be divided into regular and chaotic attractors.

The presented classification allows us to study the structure of model parameter spaces (3.3)–(3.5), which comprises the separation of spaces into regions of dynamic modes of different types and an analysis of the evolution of the chosen regions with change in model parameters.

3.3. Dynamics of Coupled Systems

We will begin our analysis of nonlinear dynamics of two cascade-connected PLLs without additional couplings with the simplest case when there are quick-acting first-order filters ($\varepsilon_1 \ll 1$, $\varepsilon_2 \ll 1$) in control circuits. Let us view the evolution of dynamic modes with increments in filters parameters ε_1 and ε_2.

3.3.1. *Case of quick-acting control circuits*

Studying dynamic processes in the ensemble with quick-acting control circuits involves an analysis of the model (3.5) moving at $\delta = \kappa = 0$. Considering the suppositions, the first generator behavior does not depend on the second generator and it follows that dynamics PLL$_1$ can be examined independently.

Dynamic processes PLL$_1$ are characterized by the equation

$$\frac{d\varphi_1}{d\tau} = \gamma_1 - \sin\varphi_1, \tag{3.6}$$

which is defined on the phase circle $T_1 = \{\varphi_1(\mathrm{mod}\,2\pi)\}$. At $|\gamma_1| < 1$ on the circle T_1, there are two states of equilibrium: stable $\varphi_1^1 = \arcsin\gamma_1$ and unstable $\varphi_1^2 = \pi - \arcsin\gamma_1$. At $|\gamma_1| = 1$, the bifurcation of stable and unstable states of equilibrium is merged at point $|\varphi_0^*| = \pi/2$, and at $|\gamma_1| > 1$, there is no state of equilibrium — the representative point which characterizes the system state moves around the circle T_1. Hence, at $|\gamma_1| \leq 1$ in the PLL$_1$ system, a synchronization mode is always realized, and at $|\gamma_1| > 1$, a regular mode of beatings is realized.

Dynamic processes of PLL_2 will be examined first for the model (3.1) by considering in ψ_2 phases the difference in signal $S_2(t)$ at the PLL_2 outlet and signal $S_1(t)$ at the PLL_1 outlet as a variable. In this case the behavior of PLL_2 is written as

$$\frac{d\psi_2}{d\tau} = \gamma_2 - \sin\psi_2 - \frac{d\varphi_1}{d\tau}. \tag{3.7}$$

At $|\gamma_1| < 1$ in PLL_1, with increase in time, a synchronization mode is set where $d\varphi_1/d\tau = 0$. In this case, Eq. (3.7) with an accuracy to table symbols coincides with Eq. (3.6), that is, the varying dynamic PLL_2 relates to the reference signal $S_0(t)$. At $|\gamma_2| < 1$ in PLL_2, signal synchronization modes $S_1(t)$ and $S_0(t)$ with an accuracy to the constant $\psi_2^1 = \arcsin\gamma_2$ is set, and at $|\gamma_2| > 1$ a mode of beatings is realized where the current difference of phases ψ_2 grows if $\gamma_2 > 1$, and decays if $\gamma_2 < -1$. At $|\gamma_1| > 1$ in PLL_1, a mode of beatings is set where φ_1 grows if $\gamma_1 > 1$, and decays if $\gamma_1 < -1$. In this case, Eq. (3.7) becomes non-autonomous. Considering that a non-autonomous equation has no state of equilibrium, it follows that in PLL_2 system there are no modes of synchronization.

By realizing the dynamic PLL_1 the relating to $S_0(t)$ and PLL_2 relating to $S_1(t)$, and taking into consideration the connection $\varphi_2 = \varphi_1 + \psi_2$, it is easy to reconstruct the dynamic PLL_2 relating to $S_0(t)$. If PLL_1 is in a synchronous mode ($|\gamma_1| < 1$), then at and $|\gamma_2| < 1$ in PLL_2 a synchronous mode is also set, now with a mistake $\varphi_2^1 = \arcsin\gamma_2 + \arcsin\gamma_1$, and at $|\gamma_2| > 1$ a mode of beatings is realized. If PLL_1 is in a mode of beatings ($|\gamma_1| > 1$), then PLL_2 relating to S_0 can be either in a mode of beatings or in a quasisynchronous mode. In a quasisynchronous mode, PLL_2 will function when the phase difference incursion φ_1 in one direction compensates the phase difference incursion φ_2 in a backward direction, that is, it is possible at $\gamma_1\gamma_2 < 0$.

All these conclusions can be obtained straight from the analysis of the system (3.5) at $\kappa = \delta = 0$. The results of this analysis are shown in Fig. 3.3. At the parameter point from the region $D_S = \{-1 < \gamma_1 < 1, -1 < \gamma_2 < 1\}$, the system (3.5) has four states of equilibrium

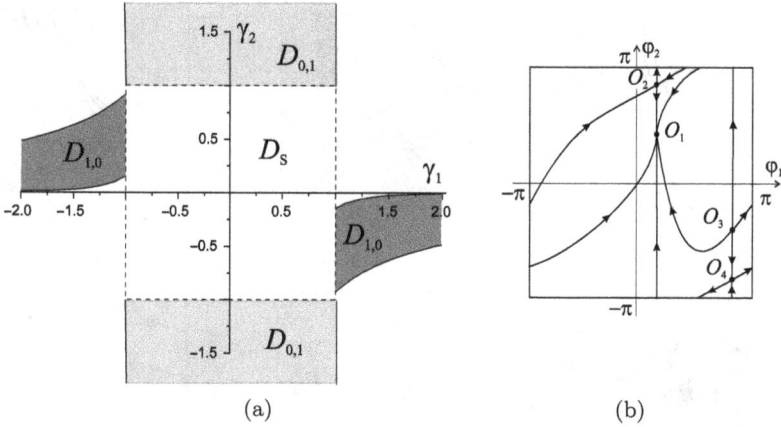

Fig. 3.3 (a) Parametric portrait of model (3.5) at $\kappa = \delta = 0$, (b) phase portrait of model (3.5) at parameter points from the region D_S.

on the torus T:

$$O_1(\varphi_1^*, (\varphi_1^* + \psi_2^*)) \quad \text{stable node,}$$

$$O_2(\varphi_1^*, \pi + (\varphi_1^* - \psi_2^*)) \quad \text{saddle,}$$

$$O_3(\pi - \varphi_1^*, -(\varphi_1^* + \psi_2^*)) \quad \text{unstable node,}$$

$$O_4(\pi - \varphi_1^*, \pi - (\varphi_1^* - \psi_2^*)) \quad \text{saddle,} \qquad (3.8)$$

where $\varphi_1^* = \arcsin \gamma_1, \psi_2^* = \arcsin \gamma_2$. The state of equilibrium O_1 is globally asymptomatically stable, and the phase portrait of the system (3.5) for the region D_S is shown in Fig. 3.3(b). Outward the region D_S on the torus T, there are periodic or quasiperiodic motions. While leaving the region D_S over the borders $\gamma_2 = 1$ and $\gamma_2 = -1$, the states of equilibrium O_1 and O_2, O_3 and O_4 merge with generation of separatrix loops of saddle-nodes O_{12}, O_{34}, which at $|\gamma_2| > 1$ are destroyed, constructing stable $L_{0,1}$ and unstable $\Gamma_{0,1}$ limit cycles that encircle the torus T only in the direction φ_2. The distinctive feature of cycles $L_{0,1}$ and $\Gamma_{0,1}$ is that the coordinate φ_1 on these cycles does not change, that is, limit cycles are lines (Fig. 3.4(a), parallel to the axis of ordinate. Considering that the projection $L_{0,1}$ on the partial subspace T_1 degenerates into a point, then this limit

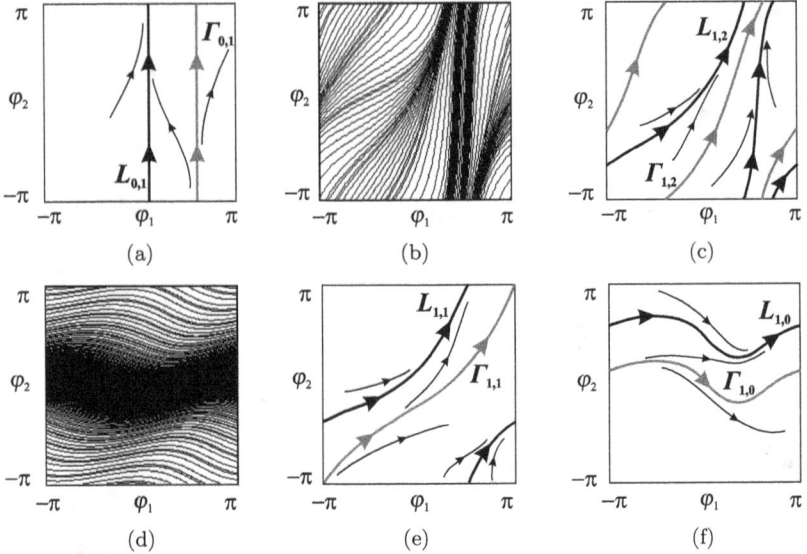

Fig. 3.4   Examples of phase portraits of model (3.5) outward the region D_S.

cycle corresponds to the partial synchronization mode of the ensemble when PLL_1 is in a synchronous mode and PLL_2 in the mode of beatings. Cycles $L_{0,1}$ and $\Gamma_{0,1}$ exist at the parameter points from the region $D_{0,1} = \{-1 < \gamma_1 < 1, |\gamma_2| > 1\}$. At the outlet from the region $D_{0,1}$ over the borders $\gamma_1 = 1$ and $\gamma_1 = -1$, cycles $L_{0,1}$ and $\Gamma_{0,1}$ merge and disappear; on the phase torus T, quasiperiodic motions appear which at variations of parameters can become periodic with different rational rotation numbers. Figure 3.4 shows examples of phase portraits of model (3.5) outward the region D_S. While leaving the region D_S over borders $\gamma_1 = 1$ and $\gamma_1 = -1$, the states of equilibrium O_1 and O_4, O_3 and O_2 merge and disappear. As a result on the torus T, periodic and quasiperiodic motions appear. At the parameter points from the region $D_{1,0}$ on the phase torus T, there are stable $L_{1,0}$ and unstable $\Gamma_{1,0}$ limit cycles which cover the torus only in the direction φ_1 (Fig. 3.4(f)). Considering that the projection of the limit cycle $L_{0,1}$ on the partial subspace $T_2 = \{\varphi_2 (\mathrm{mod}\, 2\pi)\}$ is bounded and does not degenerate into a point, it defines a quasisynchronous mode PLL_2 relating to a reference signal $S_0(t)$ and region $D_{1,0}$ is the region of a regular quasisynchronous mode with the presence of PLL_2. While leaving the region $D_{1,0}$ with growing (reducing) parameter γ_2, the cycles $L_{1,0}$ and $\Gamma_{1,0}$ merge and disappear. As a result, on the phase torus T, periodic and quasiperiodic motions appear encircling the torus in the direction φ_1 as well as in the direction φ_2, which correspond to regular modes of beatings PLL_1 and PLL_2.

So, as a result of the cascade connection of two PLLs with low-inertia control circuits in the PLL_2 system, it is possible that a new unusual partial PLL systems' with the first-order filters of a regular quasisynchronous mode will appear. A quasisynchronous mode in the PLL_2 system is possible only if PLL_1 functions in the mode of beatings.

3.3.2. *Influence of filter parameter*

In the absence of an additional connection, "backward" dynamics of PLL_1 in the ensemble is autonomous, and the influence of filters on its dynamics is described in Sec. 3.1. The role of filters in the controlling ring PLL_2 will be examined in the example of the model (3.4) in case $\varepsilon_1 = 0$. At $|\gamma_1| < 1$, the dynamics PLL_2 is equivalent to the dynamics of the autonomous PLL. At $|\gamma_1| > 1$, the model (3.4) can be described as a non-autonomous model PLL with the first-order filters. It is known [31] that such systems allow the existence of a great variety of dynamic modes as well as chaotic ones.

As an example, let us fix initial frequency mismatches PLL_1 and PLL_2 at levels $\gamma_1 = 1.03$ and $\gamma_2 = -0.1$ and analyze the development of dynamic modes of the PLL_2 system with changing inertia effects of the circuit CC_2. Evolution of PLL_2 dynamic modes at variations of parameter ε_2 in the interval $\varepsilon_2 \in (0; 65)$ is illustrated by one-parameter bifurcation diagram in Fig. 3.5.

At first, we will see what is going on in the system at growing ε_2. At $\varepsilon_2 \ll 1$ in phase space of the model (3.4), there is a periodic solution of the kind [1, 1]. It shows that both PLLs are in the mode of beatings. Projections of the stable limit cycle $L_{1,1}$, which characterizes these modes, are shown in Figs. 3.6(a) and 3.6(b). In the $\{\varepsilon_2, \varphi_2\}$ diagram, parameter points corresponding to the mode of

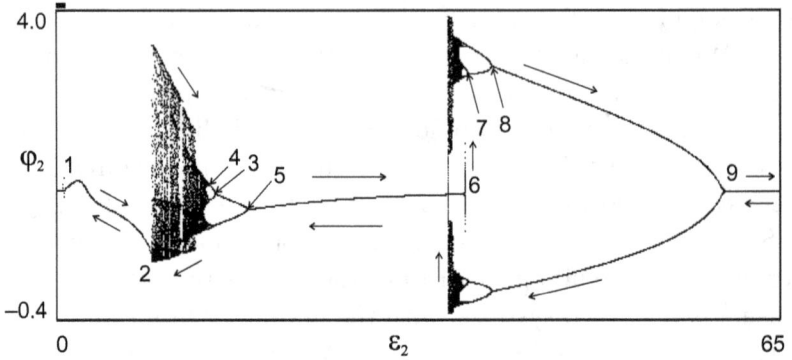

Fig. 3.5 Single-parametric Poincaré maps (3.4) bifurcation diagrams, which at $\gamma_1 = 1.03, \varepsilon_1 = 0, \gamma_2 = -0.1$.

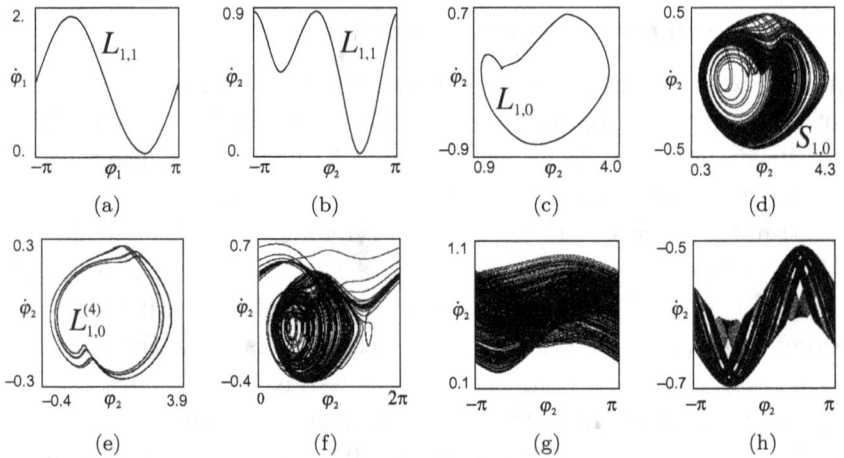

Fig. 3.6 Projections of model attractors (3.4) for $\gamma_1 = 1.03, \varepsilon_1 = 0$.

beatings PLL_2 are marked by the black line above (Fig. 3.5). When increases ε_2, dynamic processes in PLL_1 remain unchanged, but in PLL_2 they develop in accordance with the following scenario. When ε_2 starts increasing with values $\varepsilon_2 = 0.65218$, the boundary cycle $L_{1,1}$ meets with a set of period-doubling bifurcations; as a result of that in a phase space of the model (3.4), a chaotic attractor $S_{1,1}$ is of the type $[1, 1]$. In the PLL_2 system, a regular mode of beatings becomes chaotic. Then at $\varepsilon_2 = 0.6537$, attractor $S_{1,1}$ is destroyed creating a double stable limit cycle $L_{1,0}^{(2)}$ of the type $[1,0]$ (on the bifurcation diagram, it is point **1**). In the PLL_2 system, a regular quasisynchronous mode appears. By further increasing ε_2 bifurcation opposite the doubling period, at $\varepsilon = 0.6545$, cycle $L_{1,0}^{(2)}$ changes into a simple stable limit cycle $L_{1,0}$, the projection of which on the partial subspace V_2 is shown in Fig. 3.6(c). Cycle $L_{1,0}$ is kept till value of $\varepsilon = 8.625$; on the bifurcation diagram, it is point **2**. Here, cycle $L_{1,0}$ disappears as a result of a saddle-node bifurcation creating a chaotic attractor $S_{1,0}$ (Fig. 3.6). PLL_2 system comes into the mode of generation of CMO. Further increasing the parameter ε_2 leads to regularization of chaotic quasisynchronous oscillations at the outlet of PLL_2, motions of the model (3.4) return to the stable limit cycle $L_{1,0}$ (point **5**). Further increasing the parameter ε results in a multiplier ν_1 of the limit cycle $L_{1,0}$ which again turns into -1 (point **6**), but this bifurcation is now hard and the limit cycle $L_{1,0}$ disappears. Phase trajectory from its neighborhood comes onto a chaotic attractor $\bar{S}_{1,0}$ of the kind $[1, 0]$. PLL_2 system again returns into the mode of generation of CMO. Further increasing the parameter ε_2 results in the regularization of chaotic oscillations on the attractor $\bar{S}_{1,0}$ and appearance of a single stable limit cycle $\bar{L}_{1,0}$ (point **9**).

Now, let us see what happens in the system with decreasing ε_2 from $\varepsilon_2 = 0.65$. Here, a start state for creating a diagram was a limit cycle $\bar{L}_{1,0}$. When ε_2 decreases, cycle $\bar{L}_{1,0}$ goes through bifurcation of the doubling period and turns into a chaotic attractor $\bar{S}_{1,0}$. Attractor $\bar{S}_{1,0}$ at $\varepsilon_2 = 35.2$ is destroyed, and trajectory from its neighborhood directs towards the limit cycle $L_{1,0}$. When ε_2, further decreases a scenario emerges which is opposite to the scenario with increasing ε_2 as follows: *cycle $L_{1,0}$ in accordance with Feigenbaum scenario turns into a chaotic attractor $S_{1,0} \rightarrow$ attractor $S_{1,0}$ turns into cycle $L_{1,0}$ $\rightarrow$ cycle $L_{1,0}$ doubles a period becoming $L_{1,0}^{(2)} \rightarrow L_{1,0}^{(2)}$ disappears as a result of a saddle-node bifurcation, a chaotic attractor $S_{1,1}$ is set, $\rightarrow$ attractor $S_{1,1}$ turns into a cycle $L_{1,1}$.*

The information about development of the above-mentioned processes at variation of the initial frequent detuning γ_2 is obtained from

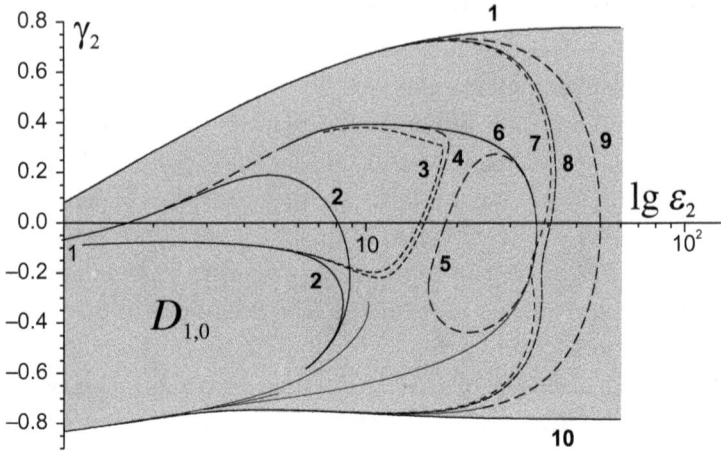

Fig. 3.7   Two-parameter model diagram (3.4) at $\gamma_1 = 1.03$, $\varepsilon_1 = 0$.

the two-parameter $\{\varepsilon_2, \gamma_2\}$-diagram in Fig. 3.7, maps of dynamic modes and maximum Lyapunov exponent are shown in Fig. 3.8. In Fig. 3.7, bifurcation curves are obtained while constructing a diagram in Fig. 3.5, which characterize the main steps in the scenario of the development of dynamic modes at change of parameter ε_2. Here, lines 1 and 10 limit the region $D_{1,0}$ of the existence of attractors of the type $[1,0]$; line 2 corresponds to the saddle-note line of cycle bifurcation $L_{1,0}$, as a result of which a chaotic attractor $S_{1,0}$ is created; lines 5, 4, 3 are responsible for the first, second and fourth periodic-doubling bifurcation of the cycle $L_{1,0}$; in line 6, saddle-node bifurcation of disappearance of the limit cycle $L_{1,0}^{(2)}$ happens; lines 9, 8, 7 are responsible for the first, second and third periodic-doubling bifurcations of the cycle $\bar{L}_{1,0}$. When leaving the region $D_{1,0}$, a global mode of beatings is realized in the ensemble. Depending on parameters, this mode can be either regular or chaotic. Examples $(\varphi_2, \dot{\varphi}_2)$ of projections of the model attractors (3.4), which characterize a chaotic mode of beatings and a double-frequency regular mode of beatings, are shown in Figs. 3.6(e) and 3.6(f), respectively. Note that attractors of the model (3.4) of types $[0,1]$ and $[1,1]$ can exist simultaneously, that is why in Fig. 3.7 the gray colored region $D_{1,0}$ is a

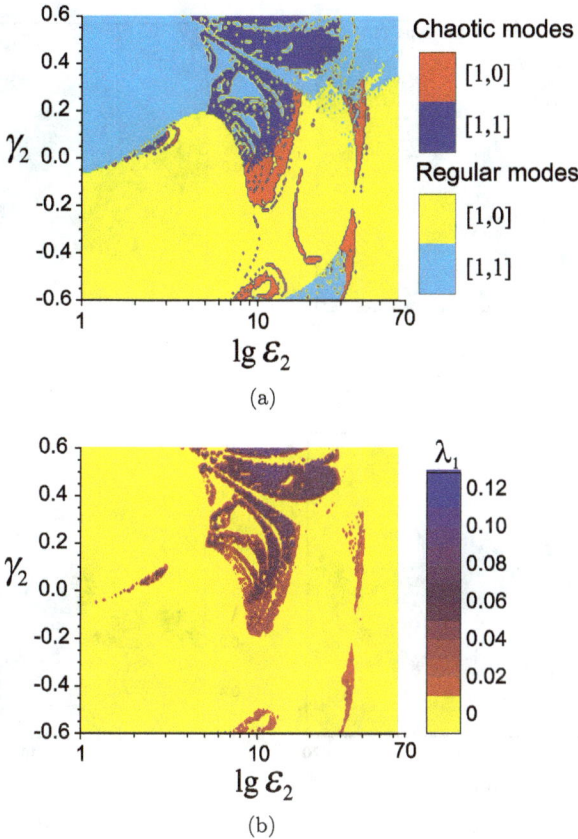

Fig. 3.8 (a) Maps of dynamic modes and (b) maximum Lyapunov exponent of model (3.4) for $\gamma_1 = 1.03, \varepsilon_1 = 0$.

region of existence of type[1, 0] modes but it is not a region of global stability.

Bifurcation diagram in Fig. 3.7 is amplified with a map of dynamic modes in Fig. 3.8. It provides insight into dimensions and structure of dynamic modes of different types of regions of existence. Total proportions of the marked out regions calculated with the help of this map have different values: region of existence of chaotic modes of the type [1, 0] occupies 5.1% of the total map area ($1 \le \varepsilon_2 \le 70, -0.6 \le \gamma_2 \le 0.6$), regions of existence of regular modes of the type [1, 0] and [1, 1] occupy 54.8% and 28.9%, respectively. Chaotic

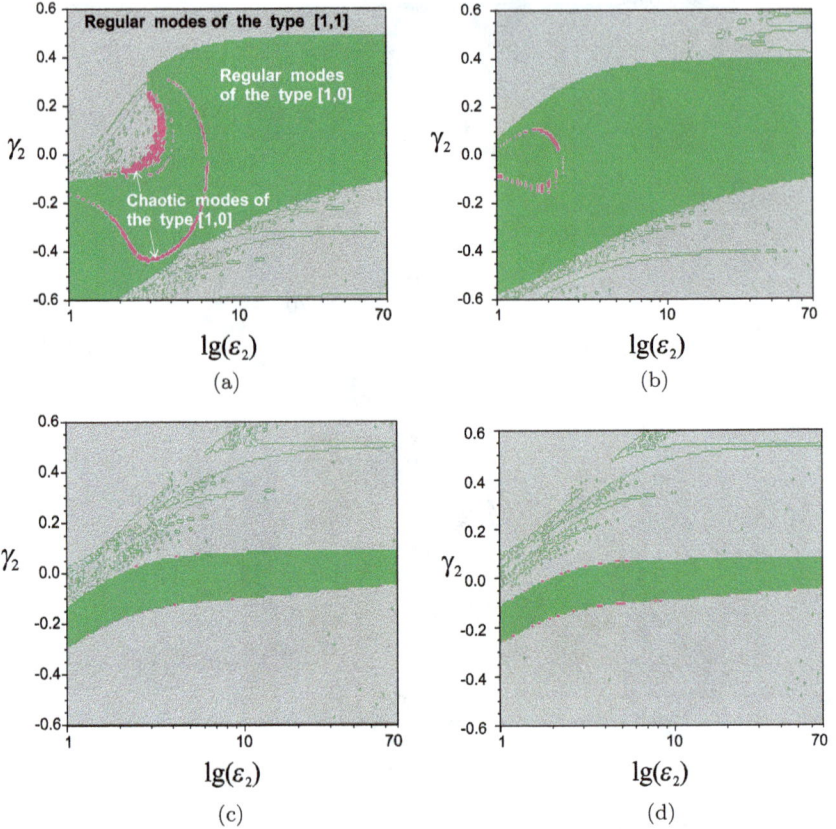

Fig. 3.9 Maps of dynamic modes of model (3.4) for (a) $\gamma_1 = 1.03$, $\varepsilon_1 = 1$, (b) $\gamma_1 = 1.1$, $\varepsilon_1 = 1$, (c) $\gamma_1 = 1.03$, $\varepsilon_1 = 5$ and (d) $\gamma_1 = 1.1$, $\varepsilon_1 = 5$.

modes are characterized by the map of maximal Lyapunov index in Fig. 3.8(b).

Previously we have examined the dynamic modes of the model (3.4) at $\varepsilon_1 = 0$, i.e., when there is no inertia effect of PLL$_1$. Now, let us examine the inertia effect of PLL$_1$ influence, and introduce a parameter $\varepsilon_1 \neq 0$. In Fig. 3.9, the results of such an analysis illustrate the maps of dynamic modes for different ε_1, and in Table 3.1 the characteristics of the received regions are shown. Using them, we conclude that the introduction of ε_1 considerably changes the map

Table 3.1 Total dimensions of regions of existence of dynamic modes PLL_2 depending on PLL_1 parameters.

Value of parameters	Mode of beatings (%)	Quasisynchronous modes (%)	
		Regular	Chaotic
$\gamma_1=1.03$, $\varepsilon_1 = 0$	40.1	54.8	5.1
$\gamma_1=1.03$, $\varepsilon_1 = 1$	44.5	53.4	2.1
$\gamma_1=1.03$, $\varepsilon_1 = 5$	85.5	14.4	0.1
$\gamma_1=1.10$, $\varepsilon_1 = 1$	47.4	52.1	0.4
$\gamma_1=1.10$, $\varepsilon_1 = 5$	86.9	12.9	0.3

of dynamic modes, and consequently, dynamics of PLL_2 exceedingly depends on parameters of PLL_1.

Thus, the examined example of two cascade-connected PLL ensemble shows that the introduction of an inertia effect into a local control circuit PLL_2 first results in the broadening of the region of dynamic modes of the type $[1,0]$; secondly, in the chaotization of the specified dynamic modes; and finally, in the overlapping of regions with multistable behavior, which lead to hysteretic phenomena appearing at variations of parameters. Examination of the specified chaotic modes depending on parameters γ_1 and ε_1 indicates that these modes are extremely sensitive to variations of parameters of the first PLL.

3.4. Influence of Couplings on the Ensemble Dynamics

3.4.1. *Synchronous modes*

Synchronous modes of the ensemble of two cascade-connected PLLs with low-inertia control circuits are defined by the stable equilibrium state of the model (3.5). At $(\gamma_1, \gamma_2, \kappa, \delta) \in C_0$ in a phase space of the system (3.5), there are four equilibrium states (stable or unstable depending on parameters):

$$O_1(\varphi_1^*, (\varphi_1^* + \psi_2^*)), \qquad O_2(\varphi_1^*, \pi + (\varphi_1^* - \psi_2^*)),$$
$$O_3(\pi - \varphi_1^*, -(\varphi_1^* + \psi_2^*)), \quad O_4(\pi - \varphi_1^*, \pi - (\varphi_1^* - \psi_2^*)), \quad (3.9)$$

where

$$\varphi_1^* = \arcsin \frac{\gamma_1 - \kappa\gamma_2}{1 - \delta\kappa}, \quad \psi_2^* = \arcsin \frac{\gamma_2 - \delta\gamma_1}{1 - \delta\kappa}, \tag{3.10}$$

$$C_0 = \left\{ \max \left[\frac{(\kappa\delta - 1) \cdot z \cdot \text{sign}\, \delta + \gamma_2}{\delta}, (\kappa\delta - 1) \cdot z + \kappa\gamma_2 \right] \right.$$

$$\left. < \gamma_1 < \min \left[\frac{(1 - \kappa\delta) \cdot z \cdot \text{sign}\, \delta + \gamma_2}{\delta}, (1 - \kappa\delta) \cdot z + \kappa\gamma_2 \right] \right\}, \tag{3.11}$$

with $z = \text{sign}(1 - \kappa\delta)$.

Stable equilibrium states O_1, O_2, O_3 and O_4 define synchronous modes I_{S1}, I_{S2}, I_{S3} and I_{S4}, respectively. These modes are distinguished by a residual difference of phases, which is defined by values of coordinate of stable equilibrium states. Whereas at values of parameters from region C_0 on the torus T, there is always a stable equilibrium state, and C_0 coincides with the region C_s of synchronous modes. On the plane (γ_1, γ_2) at $\delta \neq 0$, $\kappa \neq 0$, region C_s has a parallelogram form with angles in the points $A(1 + \kappa, 1 + \delta)$, $B(1 - \kappa, -1 + \delta)$, $C(-1 - \kappa, -1 - \delta)$, and $D(-1 + \kappa, 1 - \delta)$ (Fig. 3.10).

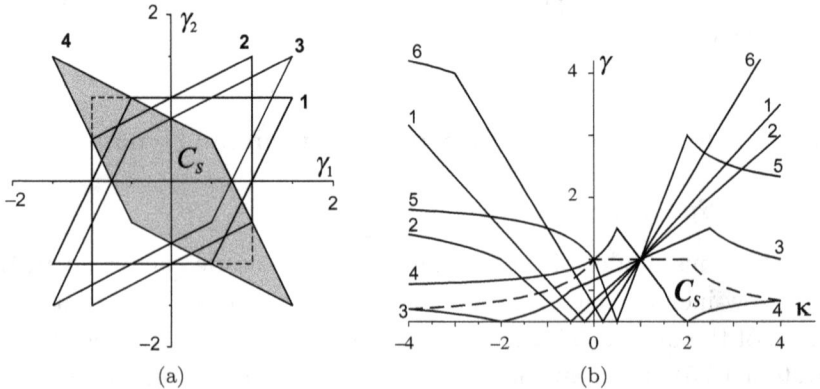

Fig. 3.10 Boundaries of the region of the model synchronous modes existence (3.5) on the plane (γ_1, γ_2) at $\delta = 0$, $\kappa = 0$ dotted line: (a) at $\delta = 0$, $\kappa = 0.5$; $\delta = 0.5$, $\kappa = 0$; $\delta = 0.5$, $\kappa = 0.5$; $\delta = 0.5$, $\kappa = -0.5$ lines 1–4; (b) on the plane (κ, γ), where $\gamma_1 = \gamma_2 = \gamma$, at $\delta = 0$ dotted line; at $\delta = -5; -2; -0.5; 0.5; 2; 5$ lines 1–6.

When there are no additional connections ($\kappa = 0, \delta = 0$), then C_s is a square such that $C_s = (|\gamma_1| < 1, |\gamma_2| < 1)$, introduction of the connection forward ($\delta \neq 0$) declines C_s vertically, introduction of the connection backward ($\kappa \neq 0$) declines C_s horizontally, and the existence of reciprocal connections ($\kappa \neq 0, \delta \neq 0$) extends C_s through one diagonal and zooms out through the other. The region of existence of equilibrium states C_s on the plane (κ, γ_1) at $\delta \neq 0$ is limited by straight lines $\gamma_1 = \kappa\gamma_2 \pm (1 - \kappa\delta)$, which cross in point $M(1/\delta, \gamma_2/\delta)$. While approaching δ to zero point, M directs forward infinite distance, and at $\delta = 0$, boundaries of the region C_s degenerate into parallel lines $\gamma_1 = \kappa\gamma_2 \pm 1$. Within the spaces of other parameters, the region of equilibrium state existence has a more complicated form. The dependence of the region boundaries C_s on the parameter δ on the plane (κ, γ) for the case $\gamma_1 = \gamma_2 = \gamma$ is shown in Fig. 3.10(b). Figure 3.10 shows that by selecting the values κ, δ, it is possible to increase the initial frequency detuning, where a synchronous mode can be established.

Stability of the model equilibrium states (3.5) is defined by signs of values

$$\Delta = 1 - \delta\kappa, \qquad \sigma = \cos \varphi_1^* \, \text{sign} \, \delta + (1 - \kappa)\cos \psi_2^*. \qquad (3.12)$$

At $\Delta > 0$, $\sigma > 0$, equilibrium state O_1 is stable; at $\Delta > 0$, $\sigma < 0$, equilibrium state O_3 is stable; at $\Delta < 0$, $\sigma < 0$, equilibrium state O_2 is stable; and at $\Delta < 0$, $\sigma > 0$, equilibrium state O_4 is stable. Change in the sign of value σ occurs on the curve

$$\gamma_s : a_1\gamma_1^2 + 2b\gamma_1\gamma_2 + a_2\gamma_2^2 = e, \qquad (3.13)$$

where $a_1 = \delta^2(1 - \kappa)^2 - 1$, $a_2 = 1 - 2\kappa$, $b = \kappa - \delta(1 - \kappa)^2$, $e = (1 - \kappa\delta)^2\kappa(\kappa - 2)$. Since $b^2 - a_1 a_2 = (1 - \kappa)^2(1 - \kappa\delta)^2 > 0$, the curve γ_s is a hyperbola. The curve γ_s corresponds either to change in the stability of equilibrium states or the conversion of the saddle value [52] into zero. In region C_s, there is always only one stable equilibrium state as equilibrium states of the model(3.5) change stability in pairs: O_1 and O_3, O_2 and O_4. At $1 < \kappa < 2$, the curve γ_s connects the points A and B, C and D (Fig. 3.11). Equilibrium state O_1 is stable

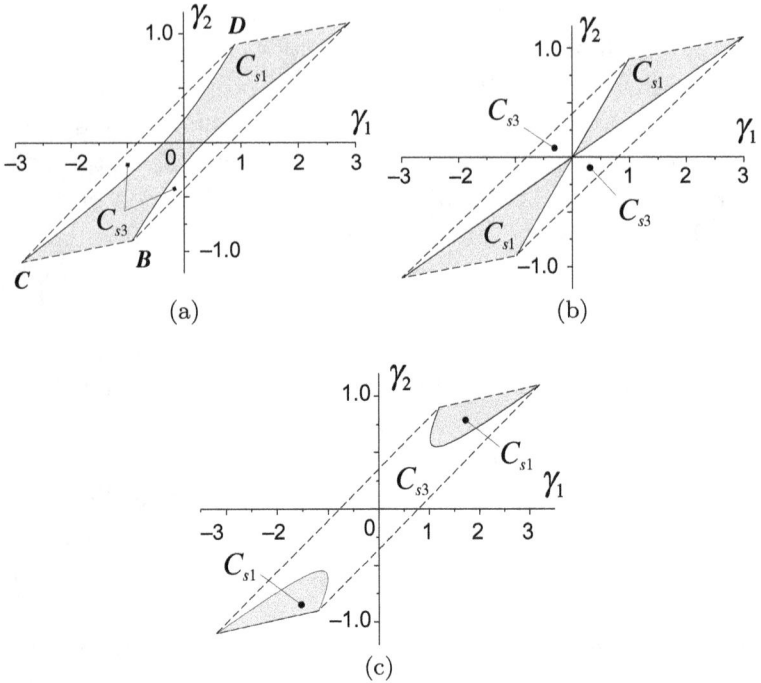

Fig. 3.11 Synchronous mode stability regions I_{s1} and I_{s3} of the model (3.5) at $\delta = 0.1$, (a) $\kappa = 1.9$, (b) $\kappa = 2$, (c) $\kappa = 2.2$.

in region C_{s1}, whereas equilibrium state O_3 is unstable. State O_3 is stable in region C_{s3}, whereas equilibrium state O_1 is unstable. At $\kappa = 2$ hyperbola γ_s degenerates into two straight lines which connect points A and C, B and D (Fig. 3.11(b)). At $\kappa > 1$, the curve γ_s connects points A and D, B and C (Fig. 3.11(c)). Now, in the major part of the region C_s there is a stable equilibrium state O_3.

Depending on a sign of Lyapunov exponent, stability loss in synchronous modes of the model (3.5) can occur either softly or hardly. Soft change in the stability of a synchronous mode I_{si} is accompanied by the appearance of a global quasisynchronous mode I_k^i, when both PLL systems function in the mode of generation of quasisynchronous oscillations. However, the appearance of quasisynchronous mode is not globally stable, simultaneously there is always a synchronous mode and other modes can also exist, e.g., a global mode of beatings. If there is a hard change in the stability of the synchronous mode, then varying scenarios in the development of dynamic processes emerge, which depend on the ensemble parameters and the initial states of the controlled generators. At hard change in the mode I_{si} in the ensemble can be installed either in the synchronous mode I_{sj}, where $j \neq i$, or in one of the modes of global or partial quasisynchronization, or global mode of beatings. The task to single out the guaranteed realization of the synchronization mode region within the space of parameters is inseparably linked with the study of self-modulated modes, provided the mechanisms for the emergence of asynchronous modes allows singling out pull-in regions into a synchronous mode with a high level of reliability.

3.4.2. *Self-oscillation modes*

Influence of additional couplings on self-oscillation modes will be examined on the example of the ensemble which unifies identical PLL systems with low-inertia control circuits, i.e., we will examine the motion of model (3.5) at $\gamma_1 = \gamma_2 = \gamma$ and $\delta = 0$. In this case, system (3.5) depends on two parameters γ and κ; it is invariant regarding the change in $\Pi_1 : (\gamma, \varphi_1, \varphi_2) \rightarrow (-\gamma, -\varphi_1, -\varphi_2)$, that is why it is enough to examine its motion in the region of positive values γ. Figure 3.12 shows that a fragment of the plane (κ, γ), where main bifurcations (3.5) are concentrated, leads to the appearance of self-oscillation modes appearing which break an ensemble's stable work into a synchronous mode.

In Fig. 3.12, the region C_s shows the existence of the equilibrium states model (3.5) with a dotted line. At parameter value from the region C_s on the surface of torus T, there are four states: O_2 and O_4 are saddles, and O_1 and O_3 either nodes or focuses. Straight line $\kappa = 2$ responds to the change in the stability of equilibrium states O_1 and O_3. To the left of this straight line (at $\kappa < 2$), equilibrium state O_1 is stable, and O_3 unstable, and to the right, the equilibrium state O_1 is unstable, and O_3 stable. Stable states O_1 and O_3 define global synchronous modes I_{S1} and I_{S3}, when both generators of the ensemble are synchronized for a reference signal. Calculating the first Lyapunov exponent showed that the boundary $\kappa = 2$ is dangerous for the mode I_{S1} and safe for I_{S3}.

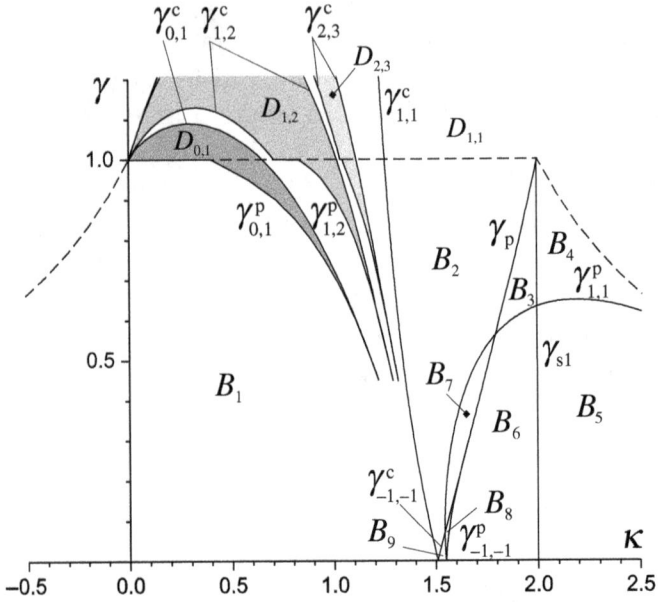

Fig. 3.12   Parametric portraits of the model (3.5) in case $\gamma_1 = \gamma_2 = \gamma$, $\delta = 0$.

Except the synchronous mode in the ensemble, quasisynchronous modes can be realized as well as modes of beatings, and at under certain conditions they can exist simultaneously. Figure 3.12 shows that curves correspond to different bifurcations of the model (3.5) and single out on the plane of parameters of (κ, γ) regions with different dynamic ensemble behavior. Here $\gamma = \gamma_{i,j}^p(\kappa)$ corresponds to the separatrix loops $\Pi_{i,j}^s$ (R_1 and R_2 of saddle O_2) and $\Pi_{i,j}^u$ (S_4 and S_3 of saddle O_4), respectively, where indices i and j characterize the number 2π which turns into homoclinic trajectories around torus T along coordinates φ_1 and φ_2, respectively. The separatrix loops $\Pi_{i,j}^s$ and $\Pi_{i,j}^u$ are stable and unstable, respectively, that is why with growing of parameters at intersection of lines $\gamma_{i,j}^p$ from loops $\Pi_{i,j}^s$ and $\Pi_{i,j}^u$, respectively, a stable $L_{i,j}$ and unstable $\Gamma_{i,j}$ limit cycles are created with the rotation number $\nu = i/j$. With further increase in parameters, cycles, $L_{i,j}$ and $\Gamma_{i,j}$ approach each other and at intersection with the curve $\gamma_{i,j}^c$ they merge and disappear. The curves $\gamma_{i,j}^p$ and $\gamma_{i,j}^c$ single out on the plane of parameters of regions $D_{i,j}$ of limit cycles

$L_{i,j}$ and $\Gamma_{i,j}$. Outside of region C_s, boundaries of regions $D_{i,j}$ are bifurcation curves $\gamma_{i,j}^c$. Note some peculiarities that characterize the complicity of structure of plane partition of model (3.5) parameters at $\kappa \neq 0$. Firstly, regions $D_{i,j}$ at $\kappa \neq 0$ penetrate into the region C_s, creating regions with bistable behavior; secondly, regions $D_{i-1,i}$ together with regions of global stability of the synchronous mode create a layer structure; while moving along the parameter κ, regions of global stability take turns with regions $D_{i-1,i}, i = 1, 2, 3, \dots$ and with increasing i store to the curve $\gamma_{1,1}^c$. (Here, "store" means the process at which distance from the curve $D_{i-1,i}$ is reduced, and regions $D_{i-1,i}$ and the regions which divide them become narrower.)

The curve $\gamma = \gamma_{1,1}^p(\kappa)$ corresponds to the separatrix loops $\widetilde{\Pi}_s$ (R_1 and R_4 of saddle O_2) and $\widetilde{\Pi}_u$ (S_2 and S_3 of saddle O_4) which encircle the torus T in directions φ_1 and φ_2 one time. The separatrix loop $\widetilde{\Pi}_s$ is stable, but $\widetilde{\Pi}_u$ is unstable. At curve crossing $\gamma = \gamma_p(\kappa)$ bottom-upwards or from right to the left on the torus T, stable $L_{1,1}$ and unstable $\Gamma_{1,1}$ limit cycles with rotation number $\nu = 1/1$ (Fig. 3.13b) are created. Coordinates φ_1 and φ_2 on the limit cycles $L_{1,1}$ and $\Gamma_{1,1}$ grow constantly. With further decrease in the parameter κ, cycles come close and at curve crossing $\gamma_{1,1}^c$ they merge and disappear.

The curve $\gamma = \gamma_{-1}^p(\kappa)$ corresponds to the separatrix loops $\widetilde{\Pi}_s^-$ (R_3 and R_2 of saddle O_2) and $\widetilde{\Pi}_u^-$ (S_1 and S_4 of saddle O_4) which encircle the torus T in directions φ_1 and φ_2 one time. The separatrix loop $\widetilde{\Pi}_s^-$ is stable, but $\widetilde{\Pi}_u^-$ is unstable. At curve crossing $\gamma = \gamma_{1,1}^p(\kappa)$ bottom-upwards or from right to the left on the torus T, stable $L_{1,1}^-$ and unstable $\Gamma_{1,1}^-$ limit cycles with rotation number $\nu = -1/-1$ are created. Coordinates φ_1 and φ_2 on the limit cycles $L_{1,1}^-$ and $\Gamma_{1,1}^-$ grow constantly. With further decrease in the parameter κ, cycles come close and at curve crossing γ_{-1}^c they merge and disappear.

The curve $\gamma = \gamma_p(\kappa)$ corresponds to the generation of homoclinic trajectories which do not encircle the phase torus.[1] Therefore, the separatrix loops of the saddles O_2 and O_4 are stable and unstable,

---

[1]In the region of coexistence of the curves γ_p and $\gamma_{1,1}^p$, the curve γ_p goes to the left of the curve $\gamma_{-1,-1}^p$ with the exception of the values $\gamma = 0$, where the curves γ_p, $\gamma_{1,1}^p$ and $\gamma_{-1,-1}^p$ merge to single point.

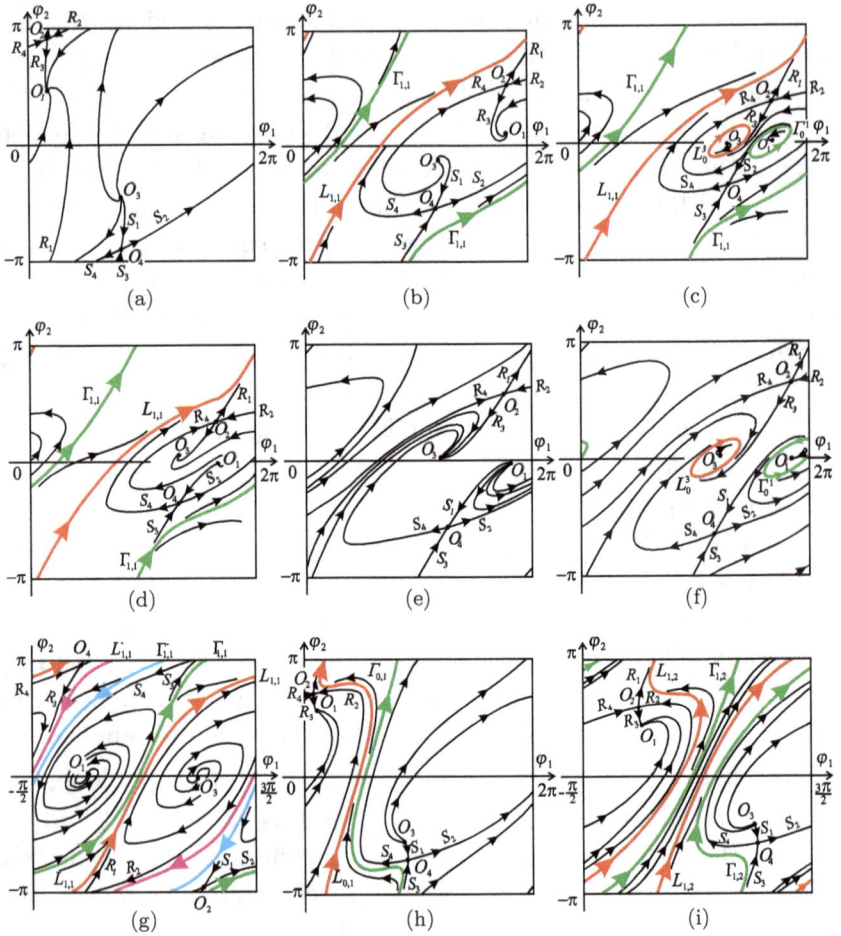

Fig. 3.13 Phase portraits of the model (3.5) for parameter points from the regions (a) B_1, B_7, (b) B_2, (c) B_3, (d) B_4, (e) B_5, (f) B_6, (g) B_9, (h) $D_{0,1}$, (i) $D_{1,2}$.

respectively. At curve crossing $\gamma_p(\kappa)$ from left to the right or from top-downward on the phase torus T, oscillatory limit cycles appear: unstable Γ_o^1 around the stable equilibrium state O_1 and unstable L_o^3 the unstable equilibrium state O_3. At increasing κ, cycle dimensions become less and at $\kappa = 2$ limit cycles collapse.

At parameter points from the region $D_{z1} = B_1 \cup B_7$ and also from the region $D_{z2} = B_5$, the system (3.5) does not have any limit

cycles. The single attractors of the model are in equilibrium state O_1
(Fig. 3.13(a)) or O_3 (Fig. 3.13(e)), which defines synchronous mode
I_{S1} or I_{S2}, respectively. The regions D_{z1} and D_{z2} are the regions of
ensemble generator's pull-in to synchronous modes.

In the region B_6 in the model (3.5) phase space together with
stable equilibrium state O_1, there is a stable oscillatory cycle L_0^3
(Fig. 3.13(f)). The cycle L_0^3 defines the mode I_k^3, when both generators function in a quasisynchronous mode. The region B_6 is a region
of bistable behavior. Here, depending on initial condition, generators
come either on the synchronous mode I_{S1} or on the quasisynchronous
mode I_k^3. In Fig. 3.12, regions of bistable behavior of the ensemble are
B_2, B_4, B_8, B_9 and $D_{i-1,i} \cap C_s, i = 1, 2, 3, \ldots$. In these regions with
the exception of the region $D_{0,1} \cap C_s$, synchronous mode I_{S1} (I_{S3})
exists together with asynchronous mode, that is why depending on
initial condition in an ensemble either a synchronous mode or a mode
of beatings is realized. The number of such limit cycle covering the
phase torus itself in the direction φ_1, as well as in the direction φ_2
(Figs. 3.13(b), 3.13(d), 3.13(f), 3.13(i)), in the regions B_8 and B_9 is
equal to two (Fig. 3.13(f)).

In the region $D_{0,1}$, there is a stable limit cycle $L_{0,1}$ encircles the
torus T only in the direction φ_2 (Fig. 3.13(h)). Projection of this
limit cycle on the partial subspace T_1 is limited by φ_1 and does not
degenerate into a point. That is why the cycle $L_{0,1}$ is responsible
for setting a mode I_{k1} of partial quasisynchronization of the ensemble, when PLL$_1$ functions in the quasisynchronous mode, but PLL$_2$
functions in the mode of beatings. So, in the subregion C_{k1} the synchronous mode I_{S1} exists together with the mode I_{k1}. In the region
$G_{k1}^z = D_{0,1} \backslash C_{k1}$, the mode I_{k1} is single and, consequently, realized
at any initial conditions. Region G_{k1}^z is a region pull-in of the PLL$_1$
system into a quasisynchronous mode I_{k1}.

A special attention should be paid to a quasisynchronous oscillation generation mechanism I_{k1}, as starting of such kind of oscillations
is, firstly, only a consequence of an additional coupling, and secondly,
is not connected with bifurcations of special trajectories in the phase
space T. At $|\gamma_1| < 1, \gamma_2 > 1, \kappa = 0$ in the ensemble, there is a par-
tial synchronous mode in which PLL$_1$ is in a synchronous mode and

PLL$_2$ is in the mode of beatings. In a phase space, a stable limit cycle in the form of a straight line responds to this mode (Fig. 3.4(a)). The mode of beatings PLL$_2$ is characterized by modulated oscillations in its local control circuit, while in the PLL$_1$ control circuit they do not exist. Organization of an additional coupling "backward" $\kappa \neq 0$ is equal to the modulation of oscillations in the PLL$_1$ control circuit with the help of a signal from the PLL$_2$ control circuit. In this case, coupling parameter κ regulates the modulation depth. So, additional coupling introduction results in starting of modulated oscillations in the system which was in a synchronous mode before. Obviously, there are always such values κ, which do not lead to the PLL$_1$ system on the mode of beatings. In the phase space T, introduction of the coupling coefficient $\kappa \neq 0$ results in "bending" of the limit cycle $L_{0,1}$. In this case, the limit cycle multiplicator does not turn into a unity and, consequently, the limit cycle $L_{0,1}$ does not have bifurcations.

In the region B_3, the ensemble demonstrates a multistable behavior. Here, depending on initial conditions, a mode of beatings can be realized together with the modes I_{S1} and I_k, which is defined by the cycle $L_{1,1}$ (Fig. 3.13(c)).

So, the performed analysis of the model (3.5) in the case $\delta = 0$, $\kappa \neq 0$ allows us to distinguish the following features of the ensemble dynamics due to an additional coupling "backward" introduction.

- Existence of two synchronous modes I_{S1} and I_{S3}, which are distinguished by synchronization errors. Transfer from the mode I_{S1} to I_{S3} occurs with increase in coupling. This transfer is hard and accompanied by hysteresis phenomenon. In a cascade system, there is a parameter range $\kappa_{z1} < \kappa < \kappa_{z2}$. When a guaranteed pull-in to a synchronous mode is impossible, maximum frequency mismatch γ, at which this pull-in is possible, does not exceed a unity.

- Existence of two quasisynchronous modes I_k^3 and I_{k1}. In the mode I_{k1}, PLL$_1$ functions in a quasisynchronous mode and PLL$_2$ functions in a mode of beatings. At $\gamma < 1$, this mode exists together with a synchronous mode I_{S1}, at $\gamma > 1$. If the mode I_{k1} exists, then it is a single ensemble dynamic mode, that is, it is realized at any initial states of PLL$_1$ and PLL$_2$. In the mode I_k^3, both PLLs function in a quasisynchronous mode. This mode exists together with the mode I_{S1}.

- Appearance of parameter regions with multistable behaviors where the final result of the ensemble functioning depends on initial states of PLL_1 and PLL_2.

- Attractors characterizing new dynamic modes appear as a result of Andronov–Hopf bifurcations, bifurcations of separatrix loops of saddle states of equilibrium, saddle-node bifurcations, oscillatory, oscillatory–rotatory limit cycles. It is necessary to mark a soft start of a quasisynchronous mode as a result of changing a limit cycle shape. This mechanism of oscillations starting in ensemble generators is not connected with bifurcations of special trajectories of the model (3.5).

3.5. Ensemble Dynamics with the First-Order Filters

Analysis of a collective dynamics of two cascade-connected PLL systems with low-inertia control circuits, mentioned in [54–57], showed a possibility of existence of varied stock of complicated dynamic modes. It is conditioned not only by couplings but also features of a mathematical model of an examined collective system, which are connected with the phase space toroidicity. In addition to this, in spite of complicity, dynamic modes of the system (3.5) are only regular, as the dimension of the phase space T is equal to two. It is natural to expect that increasing the model dimension of the examined ensemble will result in the appearance of new dynamic properties, which are connected with, for example, chaotic dynamics, while maintaining dynamic features of the low-dimension model (3.5). In this section, features of a collective behavior of the ensemble of two cascade-connected PLLs with the first-order filters with additional couplings are shown. It is important to note that individual dynamics of PLL unified in an ensemble can be only regular.

3.5.1. *Bifurcation diagram*

Influence of filter parameters on self-modulation modes of two cascade-connected PLLs, when additional couplings are deficient, was analyzed in Sec. 3.3.2. Filters' role in the ensemble with additional couplings will be examined by the example of an ensemble in which unified systems have similar initial frequency detunings $\gamma_1 = \gamma_2 = \gamma$ and parameters of control circuits $m_1 = m_2 = m$, $\varepsilon_1 = \varepsilon_2 = \varepsilon$. We will fix parameter points δ, γ, m and examine a bifurcation diagram of modes on the plane of parameters (ε, κ). Figure 3.14(a) qualitatively shows a structure of the model (3.4) parameter subspace at $\varepsilon = 0$, $\delta = 0.1$, $\gamma = 0.7$. Here, vertical chain-dotted lines limit the region

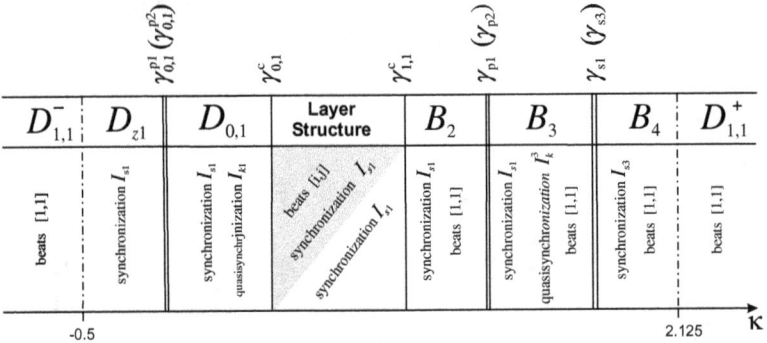

(a)

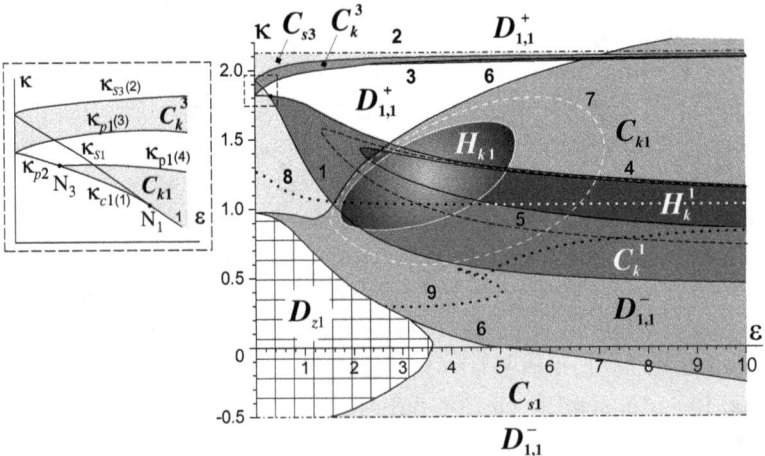

(b)

Fig. 3.14 Bifurcation diagrams of dynamic modes of the model (3.4) in case $\delta = 0.1$, $\gamma_1 = \gamma_2 = 0.7$, $m_1 = m_2 = 0.1$, (a) $\varepsilon_1 = \varepsilon_2 = \varepsilon = 0$; (b) $\varepsilon > 0$.

of existence of the synchronous modes C_s, unbroken lines divide a parameter space into regions with different dynamic behaviors, double lines characterize the boundaries where two bifurcations occur simultaneously.

Figure 3.14(b) shows a bifurcation diagram which illustrates the evolution of a parameter space structure and dynamic modes of the model (3.4) at changing values of the parameter ε from 0 to 10. Here, with the help of chain-dotted lines, a region $C_0 = \{\frac{1-\gamma}{\delta-\gamma} < \kappa < \frac{1+\gamma}{\delta+\gamma}\}$ of equilibrium state existence is singled out: $C_{S1} = \{\frac{1-\gamma}{\delta-\gamma} < \kappa < \kappa_{s1}\}$ where there is a synchronous mode I_{s1}, defined by the stable equilibrium state O_1; $C_{S3} = \{\kappa_{s3} < \kappa < \frac{1+\gamma}{\delta+\gamma}\}$ with the synchronous mode I_{s3}, defined by the stable equilibrium state O_3; and regions $C_u = \{\kappa_{s1} < \kappa < \kappa_{s3}\}$, where there are no synchronous modes as all equilibrium states of the system (3.4) are unstable. The curves $\kappa = \kappa_{s1}(\gamma, \delta, \varepsilon, m)$ (curve 1) and $\kappa = \kappa_{s3}(\gamma, \delta, \varepsilon, m)$ (curve 2) respond change in to stability from O_1 and O_3.

Calculation of the first Lyapunov exponent on the curve κ_{s3} performed according to [58], showed that this curve is safe. At its crossing with decreasing κ in the surroundings of equilibrium states O_3, a stable oscillatory limit cycle L_o^3 (Fig. 3.15(a)) emerges softly. When

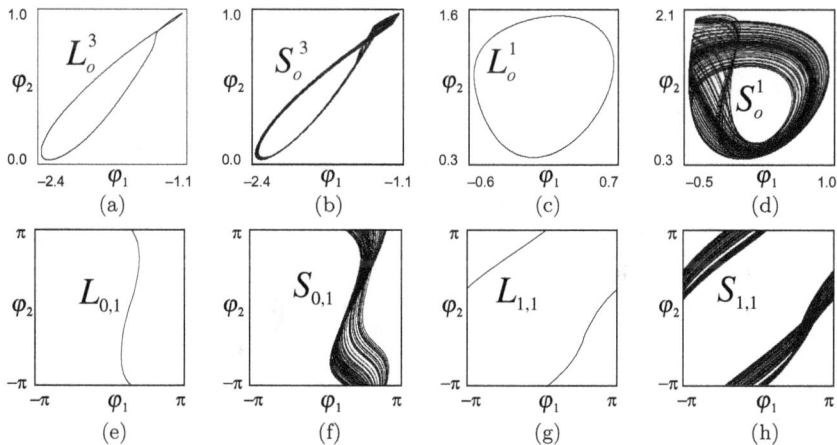

Fig. 3.15 Examples of the projections of the model attractors (3.4) which characterize the main dynamic modes of the ensemble of two cascade-connected PLL with the first-order filters.

κ decreases, cycle amplitude L_o^3 grows and at parameter points on the curve $\kappa = \kappa_{p1}(\gamma, \delta, \varepsilon, m)$ (line **3**), the cycle merges into a stable separatrix loop of the saddle O_2 which does not encircle the cylinder V. The cycle L_0^3 defines the quasisynchronous mode I_k^3, which exists at region $C_k^3 = \{\kappa_{p1} < \kappa < \kappa_{S3}\}$. It is remarkable that the cycle L_o^3 can lose stability through period-doubling bifurcation and as a result of that on its base a chaotic oscillatory attractor S_o^3 (Fig. 3.15(b)) is formed. Attractor S_o^3 corresponds to the generation of CMO at the outlet of both ensemble generators. In the examined case, the parameter regions where CMO exist are extremely small, that is why in Fig. 3.14 they are not marked.

The first Lyapunov exponent on the curve κ_{s1} changes the sign in the point N_1 (see the fragment in Fig. 3.14). To the left of the point N_1, it is positive; here the change in the state stability O_1 occurs at crossing κ_{s1} with increase in the parameter κ or ε as a result of an unstable oscillatory cycle Γ_o^1 constricting into it. The cycle Γ_o^1 emerges either from the separatrix loop of the saddle O_4 at crossing bottom-up (from left to the right) of the curve $\kappa = \kappa_{p2}(\gamma, \delta, \varepsilon, m)$, or as a result of a saddle-node bifurcation on the curve $\kappa = \kappa_{c1}(\gamma, \delta, \varepsilon, m)$. At curve crossing κ_{c1} bottom-up together with the cycle Γ_o^1 a stable oscillatory cycle L_o^1 emerges (Fig. 3.15(c)), which at further increase in κ disappears into a separatrix loop of the saddle O_2 on the curve κ_{p1} (line **4**, Fig. 3.14). Oscillatory attractor around the equilibrium state O_1 exists at parameter points from the region C_k^1, which is situated between lines **1** and **4**. Line **1** includes a bifurcation curve κ_{c1} and a part of the curve κ_{s1}, which is situated to the right of the point N_1. Line **4** includes a part of the bifurcation curve κ_{p1}, on which a saddle value is negative, and a curve of the saddle-node bifurcation κ_{c2}. Lines **1** and **4** are joined at the point N_3, where as a result of a heteroclinic trajectory created between saddle equilibrium states O_2 and O_4, the curve κ_{p2} stops existing.

The oscillating attractor around the equilibrium state O_1 defines the global quasisynchronous mode which can be regular as well as chaotic. Chaotic oscillatory attractor S_o^1 (Fig. 3.15(d)) corresponds to the chaotic quasisynchronous mode I_k^{h1} in the phase space of the model (3.4). It appears based on the limit cycle L_o^1 as a result of the cascade of period-doubling bifurcations and exists at the parameter points from the region H_k^1. In Fig. 3.14, the region H_k^1 is marked by the dark color, in its boundary, there is a curve which respond to the fourth period-doubling bifurcation of the cycle L_o^1. The dotted line 5, which goes inside the region C_k^1, corresponds to the first change in the stability in the limit cycle L_o^1.

The region C_{k1} is the region of existence of attractors of the type [0, 1] (encircling the phase cylinder V only in the direction φ_2). These attractors correspond to the partial quasisynchronization mode, while the first generator of the ensemble is in the quasisynchronous mode with respect to the reference signal, and the second generator in the mode of beatings. It is established that the partial synchronization mode can be regular as well as chaotic. Examples of the attractors projections which characterize regular and chaotic modes of the partial quasisynchronization are shown respectively, in Figs. 3.15(e) and 3.15(f). A chaotic mode is formed based on the stable cycle $L_{0,1}$ through the series of periodic-doubling bifurcations and exists at parameters from the region H_{k1}. The region C_{k1} is limited in Fig. 3.14 by line **6**, which is composed of curve fragments respond to the creation of homoclinic trajectories and double limit cycles encircling the phase cylinder V only in the direction φ_2. The dotted line 7, that goes inside the region C_{k1}, responds to the first change of the stability in the limit cycle $L_{0,1}$.

The regions $D_{1,1}^+$ and $D_{1,1}^-$ are the regions of existence of attractors of the type [1, 1]. At $\varepsilon = 0$, this attractor is the limit cycle $L_{1,1}$ with the rotation number $\nu = 1/1$, the region $D_{1,1}^-$ borders with the boundary of the region C_0, and $D_{1,1}^+$ creates, in the region of existence of synchronous and quasisynchronous modes, creates, regions of bi- and multistable behaviors of the ensemble. In Fig. 3.14(b), the region $D_{1,1}^+$ is limited by the bifurcation curve $\gamma_{1,1}^c$ (line 8). By crossing this line bottom-up as a result of the saddle-node bifurcation in the phase space of the model (3.4), the limit cycle $L_{1,1}$ emerges (Fig. 3.15(g)), which is responsible for a regular global mode of beatings. The region $D_{1,1}^-$, at increasing ε, comes into the region C_0, thereby reducing the pull-in region in the synchronous mode I_{S1}. With increase in ε, the character of the boundary of the region $D_{1,1}^-$ (line **9**) is changed and it can become multivalue at the parameter κ. In the region $\varepsilon > 4$, while approaching line **9** there are period-doubling bifurcations of the cycle $L_{1,1}$, based on which a chaotic oscillatory attractor $S_{1,1}$ is formed (Fig. 3.15(h)), in the ensemble a global chaotic mode of beatings starts. Further increase in κ results in crisis of the attractor $S_{1,1}$, and the ensemble, as a rule, comes to the mode of the partial quasisynchronization.

Asynchronous modes $I_{i,j}$, defined by rotatory attractors of the system (3.4), unlimited by coordinates φ_1 and φ_2, exist in the regions $D_{i,j}$. At little time constants ε, the regions $D_{i,j}$ keep the "layer" structure of the space of parameters installed at $\varepsilon = 0$. With increase in ε, movements in regions with asynchronous modes lead to collision and the "layer" structure is destroyed, thus transforming into a region of bistable behavior, where a synchronous mode I_{s1} and different modes of beatings exist simultaneously. Further increase in ε leads to broadening of the region of existence of the modes of beatings. In this case, regular attractors which are responsible for the mode of beatings can go through bifurcations leading to their chaotization. As a result of this in the ensemble, chaotic modes of beatings of different forms appear (Fig. 3.16).

The region D_{Z1} is the region of pull-in of the cascade system into the synchronous mode I_{s1}. At initial parameter points from this region, the mode I_{s1} is globally stable, that is, it is installed at any initial terms. The boundary of the pull-in region has a complicated structure. In Fig. 3.14(b), it contains areas of the following bifurcation curves: separatrix loops of the saddle (saddle-node) encircling the cylinder V in the direction φ_2, separatrix loops encircling the cylinder V by φ_2, as well as by φ_1, saddle-node bifurcation of different rotatory limit cycles, crisis of the rotatory chaotic attractor.

So, from the offered results, it is clear that the introduction of an inertia effect into a control circuit leads to material changes in the ensemble dynamics. This results in the appearance of new parameter regions, where there are no synchronous modes, for broadening of regions of existence of quasisynchronous and asynchronous modes,

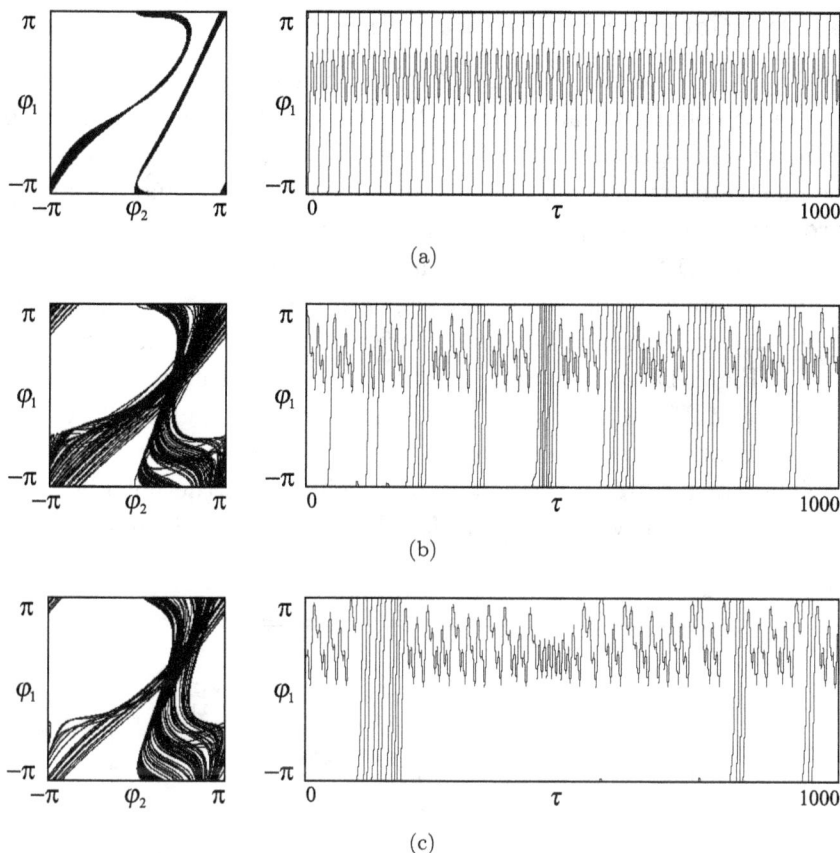

Fig. 3.16   Phase projections of rotatory attractors of the model (3.4) at (a) $\varepsilon = 3$, $\kappa = 0.97$; (b) $\varepsilon = 3, \kappa = 0.99$; (c) $\varepsilon = 3.5, \kappa = 1$.

for chaotization of dynamic modes, for decreasing of pull-in regions into a synchronous mode and making its boundary character more complicated. In the diagram $\{\varepsilon, \kappa\}$ the installation of the parameter region with different types of attractors of the model (3.4) defines the scenarios for development of generator modes change in with parameters ε and κ. We will examine some of these scenarios illustrating behavioral features of the ensemble generators with inertia control circuits at variations of coupling parameter κ.

3.5.2. *Coupling parameters influence*

We will examine the scenario of developing the ensemble dynamic modes at coupling changing κ in sections $\varepsilon = 3.0$ and $\varepsilon = 5.0$ of the bifurcation diagram in Fig. 3.14. The evolution of dynamic modes will be illustrated by one-parameter bifurcation diagrams and respective projections of phase portraits of the system attractors (3.4).

Figure 3.17 shows $\{\kappa, \varphi_1\}$-diagram of reconstruction of PLL_1 dynamic modes while leaving the region D_{z1} pull-in to a synchronous mode I_{s1} at increasing values of parameter κ till the value reaches $\kappa = 2$ (Fig. 3.17) and the return entry into D_{z1} at decreasing κ values from $\kappa = 2$ (Fig. 3.17(b)). The dark bar above the diagram illustrates rotations at the coordinate φ_1.

The start state of the model (3.4) while constructing Fig. 3.17 was an oscillatory limit cycle L_o^1 (Fig. 3.15(c)), which appeared

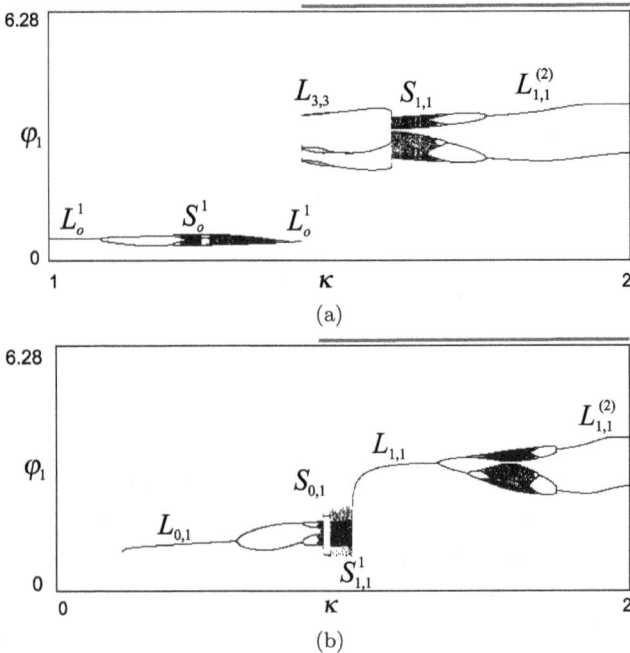

Fig. 3.17   One-parameter bifurcation diagrams of the model (3.4) at $\delta = 0.1$, $\gamma = 0.7$, $m = 0.1$, $\varepsilon = 3$.

at $k = 0.708241$ as a result of a soft change in the stability of
the equilibrium state O_1. For an ensemble, appearance of the limit
cycle L_o^1 is equivalent to changing the synchronous mode I_{s1} on
the regular quasisynchronous mode I_k^1. At increasing κ, the cycle
L_o^1 comes through a series of period-doubling bifurcations (at $\kappa =$
$1.082641; 1.212592; 1.2247573; 1.2266931$ and etc.); as a result, in the
phase space of the model (3.4), a chaotic oscillatory attractor S_o^1
(Fig. 3.15(d)) appears. In the ensemble, a mode is installed, in which
at the outlet of both ensemble generators there are chaotically modulated oscillations.

With further increase in κ, the chaotic attractor S_o^1 through bifurcations which are opposite to the period doubling, again becomes a
limit cycle L_o^1 (at $\kappa = 1.412124$). In the ensemble, a regular mode
I_k^1 is installed again. It is remarkable that the interval of existence
of the chaotic attractor S_o^1 by the parameter κ contains regular windows, in which a chaotic attractor becomes a stable multiturn cycle.
For example, in the interval $1.263746 < \kappa < 1.271138$ chaotic oscillations become regular, and a global regular quasisynchronous mode is
realized in the ensemble (Fig. 3.18(a)). At $\kappa = 1.43370$, the cycle L_o^1

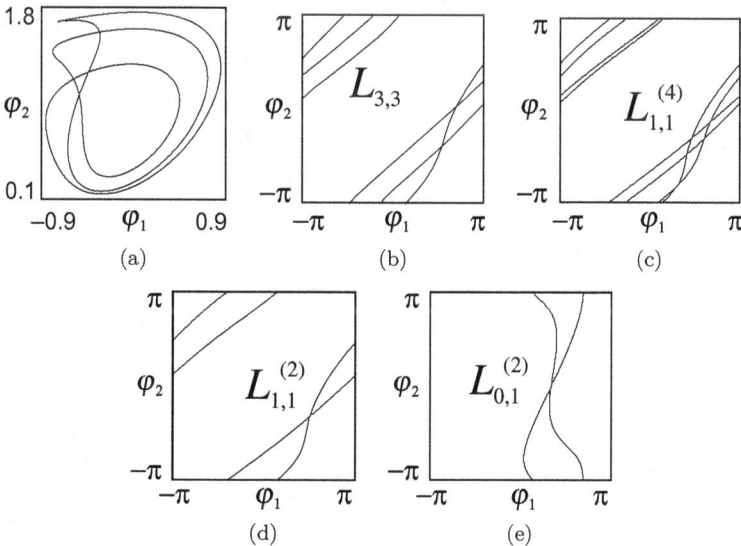

Fig. 3.18 Examples of projections of the model attractors (3.4).

disappears as a result of a saddle-node bifurcation and phase trajectories from its surroundings are attracted by the rotary limit cycle $L_{3,3}$ (Fig. 3.18(b)). In the ensemble, a regular mode of beatings is installed. At $\kappa = 1.5903$, the cycle $L_{3,3}$ disappears, phase trajectories from its surroundings are attracted by the chaotic rotatory attractor $S_{1,1}$ (Fig. 3.15(g)). In the ensemble, a chaotic mode of beatings is installed. Amplification of the coupling κ leads to regularity of the chaotic attractor $S_{1,1}$ and its further simplification: at $\kappa = 1.684$, the attractor $S_{1,1}$ turns into a rotatory limit cycle $L_{1,1}^{(4)}$ (Fig. 3.18(c)); at $\kappa = 1.752$ the attractor $S_{1,1}$ turns into $L_{1,1}^{(2)}$ (Fig. 3.18(d)); and at $\kappa = 2.397$ the attractor $S_{1,1}$ turns into $L_{1,1}$ (Fig. 3.15(g)). In the ensemble, a regular mode of beatings is installed.

Now, we will see what happens in the ensemble at decreasing κ from 2 to 0. This process is illustrated by the diagram in Fig. 3.17(b). At $\kappa = 2$, the single attractor of the system $\kappa = 2$ is a limit cycle $L_{1,1}$ (Fig. 3.15(g)). At decreasing bonding force κ, the cycle $L_{1,1}$ turns into a chaotic attractor $S_{1,1}$ (Fig. 3.15(h)), which then again becomes the limit cycle $L_{1,1}$ (Fig. 3.15(g)). At $\kappa = 1.035$, the cycle $L_{1,1}$ disappears as a result of a saddle-node bifurcation; it is changed by a rotatory chaotic attractor $S_{1,1}^1$ (Figs. 3.16(b) and 3.16(c)) and in the ensemble a regular mode of beatings is changed by a chaotic mode of beatings. At decreasing κ an internal bifurcation of the attractor $S_{1,1}^1$ occurs, as a result rotation by the coordinate φ_1 disappears on it and a rotatory chaotic attractor $S_{1,1}^1$ turns into an oscillatory–rotatory chaotic attractor $S_{0,1}$ (Fig. 3.15(f)). For the ensemble, it means that from the chaotic mode of beatings it turns to partial chaotic quasisynchronization mode, in which at the outlet of PLL$_1$ there are CMO, and PLL$_2$ continues to function in a chaotic mode of beatings. Further decrease in κ leads to regularization of oscillations, the chaotic attractor $S_{0,1}$ turns into a stable limit cycle $L_{0,1}$ (Fig. 3.15(e)). At $\kappa = 0.2267$, the cycle $L_{0,1}$ directs towards a stable equilibrium state O_1 — in the ensemble, there is a pull-in to a synchronous mode I_{s1}.

Now, we will increase the parameter ε. In Fig. 3.19, there are $\{\kappa, \varphi_1\}$-diagrams of reconstruction of dynamic modes of PLL$_1$ at the level $\varepsilon = 5$ when parameter κ increases from -0.5 to 2 (Fig. 3.19(a))

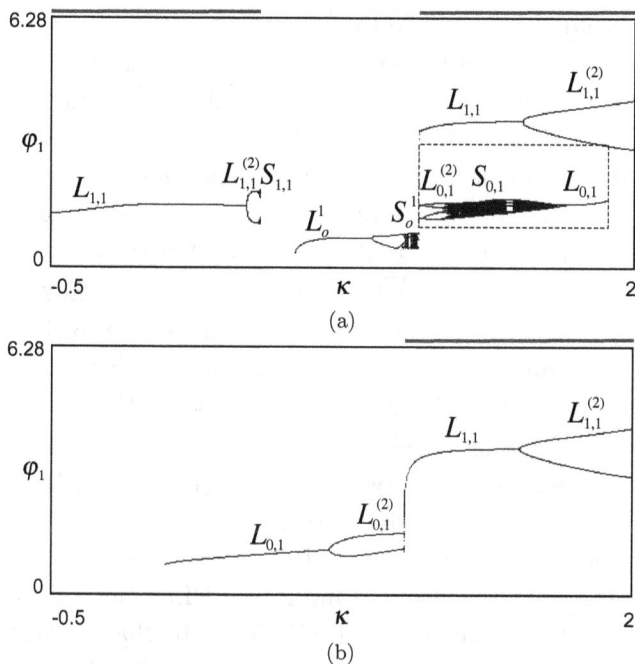

Fig. 3.19 One-parameter bifurcation diagrams of the model (3.4) at $\delta = 0.1$, $\gamma = 0.7$, $m = 0.1$, $\varepsilon = 5$.

and κ decreases from 2 to -0.5 (Fig. 3.19(b)). The start state of the model (3.4) for constructing the diagram in Fig. 3.19 was a rotatory limit cycle $L_{1,1}$ (Fig. 3.15(g)), which corresponds to the installation of a global regular mode of beatings in the ensemble.

When parameter κ increases based on the cycle $L_{1,1}$ as a result of period-doubling bifurcations, a chaotic rotatory attractor $S_{1,1}$ (Fig. 3.15(h)) is formed; the ensemble comes on the chaotic mode of beatings. At $\kappa = 0.401$, the chaotic attractor $S_{1,1}$ is destroyed, phase trajectories from its surroundings turn towards a stable equilibrium state O_1. In the ensemble, a synchronous mode I_{s1} is installed. Increase in the parameter κ leads to loss of stability in the equilibrium state O_1 (at $\kappa = 0.552$), and a stable oscillatory limit cycle L_o^1 (Fig. 3.15(c)) is created around it. In the ensemble a regular mode of global synchronization I_k^1 is installed. With increase in coupling, chaotization of quasisynchronous oscillations occurs at the outlet of both ensemble generators. In the phase space of the model (3.4), this process is accompanied by period doubling of oscillatory limit cycles and is stopped by the appearance of a chaotic oscillatory attractor S_k^1 (Fig. 3.15(d)). Further, the chaotic attractor S_k^1 is destroyed, then the development of the ensemble dynamics goes by one of the two scenarios depending on values of phase variables of the model (3.4): phase trajectories go either to the area of attraction of the rotatory limit cycle $L_{1,1}$ (Fig. 3.15(g)) or to the region of attraction of the oscillatory–rotatory limit cycle $L_{0,1}^{(2)}$ (Fig. 3.18(e)). In the first case, ensemble generators come to the regular mode of beatings, increasing κ leads to change in stability of the limit cycle $L_{1,1}$ and creation of the limit cycle $L_{1,1}^{(2)}$ (Fig. 3.18(d)), which, at further increase in κ, is again transformed into the cycle $L_{1,1}$. In the second case (a part of the bifurcation diagram reflecting the second scenario of development is marked in Fig. 3.19 by the dotted line), PLL$_1$ turns into a regular quasisynchronous mode and PLL$_2$ turns into a regular mode of beatings. Increasing κ firstly leads to chaotization of regular oscillations, when limit cycle $L_{0,1}$ turns into oscillatory–rotatory attractor $S_{0,1}$ (Fig. 3.15(f)), then by regularization of the chaotic attractor, $S_{0,1}$ becomes again the cycle $L_{0,1}$. At $\kappa = 1.88$, the cycle $L_{0,1}$ disappears as a result of a saddle-node bifurcation, phase trajectories from its surroundings turn towards a rotatory limit cycle $L_{1,1}^2$ (Fig. 3.18(d)). In the ensemble, both generators turn into a regular mode of beatings.

Now, we will see what happens in the ensemble at decreasing κ from 2 to -0.5. This process is illustrated by the bifurcation diagram in Fig. 3.19(b). At $\kappa = 2$, the ensemble generators are in the regular mode of beatings, where a limit cycle $L_{1,1}^{(2)}$ responds to (Fig. 3.18(d)). At decreasing κ, the cycle $L_{1,1}^{(2)}$ turns into the cycle $L_{1,1}$, which disappears as a result of a saddle-node bifurcation at $\kappa = 1.0297$. Phase trajectories from the surroundings of the cycle $L_{1,1}$ turn on to $L_{0,1}^{(2)}$ (Fig. 3.18(e)), PLL$_1$ turns into a mode of generation of quasisynchronous oscillations, PLL$_2$ continues functioning in a regular mode of beatings. By further decreasing κ, the cycle $L_{0,1}^{(2)}$ turns into the cycle $L_{0,1}$, which does not qualitatively change the ensemble dynamics, further at $\kappa = -0.003$ the cycle $L_{0,1}$ disappears, phase trajectories from its surroundings turn into a stable equilibrium state O_1. In the ensemble, pull-in to a synchronous mode I_{s1} occurs. Synchronous mode is kept till values of $\kappa = -0.5$, that is, till the equilibrium state O_1 exists. At $\kappa < -0.5$ in the ensemble, a regular mode of beatings is installed, defined by the limit cycle $L_{1,1}$. It is remarkable that pull-in into a synchronous mode I_{s1} at increasing and decreasing parameter κ occurs at different values of κ. When pull-in to a synchronous mode with decreasing κ occurs (at κ values less than that in the case of increasing κ), the region of guaranteed pull-in to a synchronous mode will be absent here.

3.5.3. *Chaos generation*

The examination of movements of the model (3.4) shows that the ensemble of two cascade-connected PLLs with the first-order filter demonstrates a great stock of chaotic modes. A special place among these modes is taken by chaotically modulated oscillations (CMO), whose average frequency is stabilized by a reference signal. Such oscillations can be very interesting for realization of information transmission when dynamic chaos uses them as carrier oscillations. Further examination will concentrate on studying CMO, whose mathematical images are chaotic oscillatory attractors and of oscillatory–rotatory kind [59]. It is remarkable that in the models of the connected PLLs, in contrast to a single PLL, chaotic attractors of oscillatory and also oscillatory–rotatory types correspond to CMO modes (see Sec. 3.2).

The simplest case — when in the ensemble of two cascade–connected PLLs chaotic oscillations generate — is a case when the PLL system of the first order is a PLL without a filter in the control circuit: $K(p) = 1$ and the second order PLL with an integrating filter in the control circuit: $K(p) = (1 + Tp)^{-1}$, without any additional couplings. This case is examined in detail in Sec. 3.3.2, where it is found that regions of existence of chaotic oscillatory–rotatory attractors of the model of such ensemble are not too large, and there are no attractors of the oscillatory type. Broadening the abilities of connected systems by generation essentially is possible with the introduction of a feedback from the second subsystem to the first

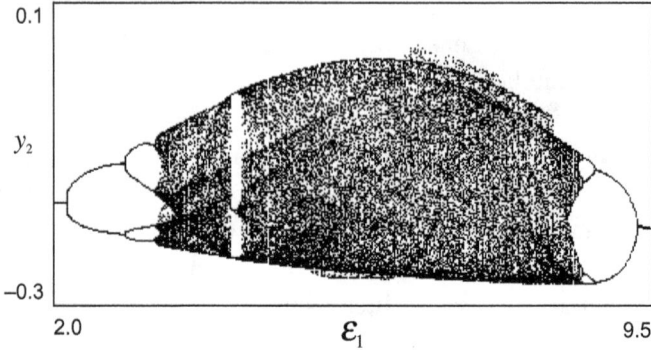

Fig. 3.20   Chaos generation through the cascade of doubling period bifurcations of oscillatory attractor in the model (3.4) at $\gamma_1 = 0.9$, $\gamma_2 = 0.7$, $\varepsilon_2 = 3$, $\kappa = 0.8$.

($\kappa \neq 0$). We will examine the most frequently realized scenarios of turning to CMO and regions of existence of these oscillations in the parameter space.

It is established that the main bifurcation mechanisms of chaotization of quasisynchronous modes in the examined ensemble are Feigenbaum scenario and the intermittency of the first kind.

In Fig. 3.20, there is a one-parameter bifurcation diagram of Poincaré mapping illustrating the process of beginning of chaotic oscillations through the cascade of period-doubling bifurcations simultaneously at the outlet of both ensemble generators, that is, a chaotic attractor is formed based on the oscillatory attractor. Here, chaotic oscillations are generated at variations of time constant of the filter in the control ring of PLL$_1$. Similar ways of chaotization of oscillatory movements of the model (3.4) can be observed at variations of other model parameters, though it is obvious that the regions of existence of CMO of the type $[0, 0]$ cannot exceed the region of existence of equilibrium states. At variations of parameters, chaotic attractors can again become regular, as in Fig. 3.20, be destroyed, or undergo internal bifurcations. Internal bifurcations of the oscillatory chaotic attractor, as a rule, are connected with the fact that one of the cycle coordinates on the attractor becomes unrestrictedly increasing (decreasing). Oscillatory attractor becomes oscillatory–rotatory.

In the ensemble, it means that only one of the generators comes to the mode of beatings, the other continues to work in the mode of generation of CMO. As oscillatory and rotatory stages on the oscillatory–rotatory attractor can have different time scales, the internal bifurcation of the attractor can lead to change in the spectral characteristics of CMO, for example, to its broadening. Regions of existence of partial quasisynchronization modes in comparison with regions of existence of global quasisynchronous modes have larger dimensions, as they are not connected with regions of existence of synchronous modes. Chaotic attractors with rotation indexes $[0, 1]$ and $[1, 0]$ can appear also as a result of period-doubling bifurcations of appropriate limit cycles and through intermittency. Figure 3.21 shows the movement of the model (3.4) toward chaotic oscillations of the type $[0, 1]$ through intermittency of the first kind.

Knowledge about bifurcation mechanisms of chaotic oscillations generations allows us to single out the regions of generation of CMO of different types in the space of assembly parameters. Figures 3.22–3.24 show some parameter regions in which the ensemble can generate CMOs: at the outlet of the second ensemble generator (Fig. 3.24), at the outlet of the first generator (Fig. 3.23), at the outlet of the first and the second generators simultaneously (Fig. 3.22). While singling out, regions in the boundary bifurcation curves were used. In Figs. 3.22–3.24, areas of boundaries corresponding to their movement toward chaos through the cascade of period-doubling bifurcations are marked by number **1**, by number **2**, where internal bifurcation occurs, and by number **3**, where the attractors crisis occurs. (Here, conceptions of crisis and internal attractor bifurcations defined in [51] are used.)

Figures 3.22–3.24 help to understand the size of regions of existence of CMO and changes in dynamic behavior which occur at the boundaries of these regions. However, it is not enough for efficiency rating of the examined ensemble as a generator of CMO. Among other things, they do not contain information of transformation of chaotic movements at variations of parameters inside the marked regions, which is extremely important for rule-making for the practical use of chaotic oscillations. Analysis of the internal structure of the

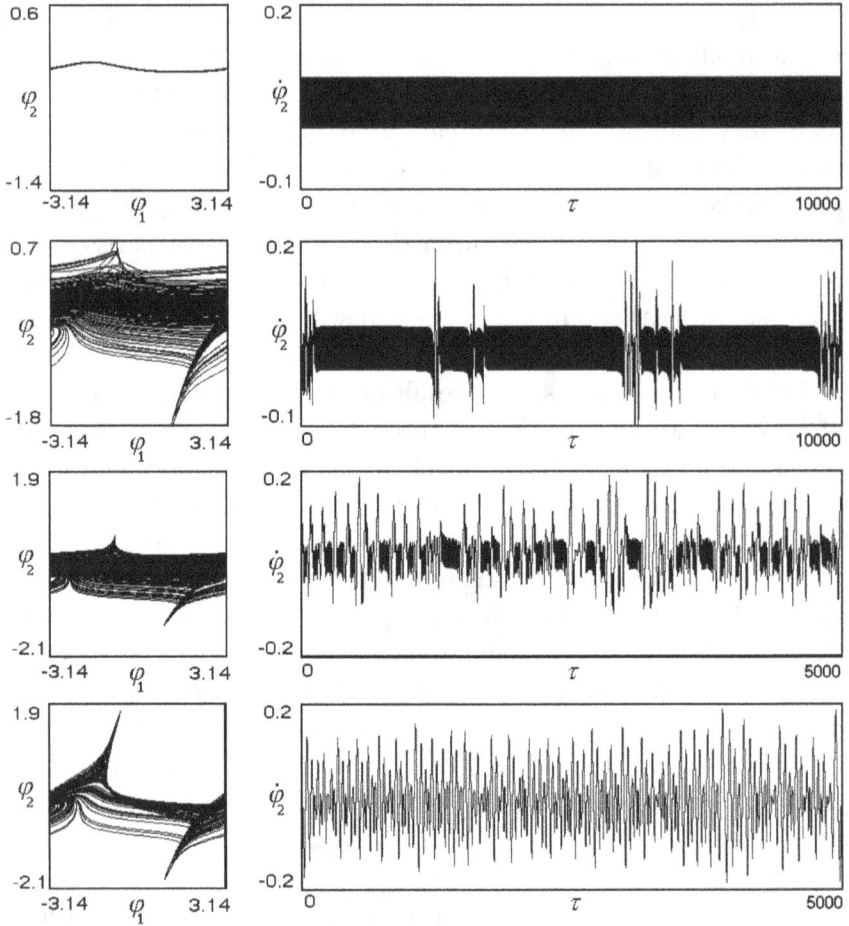

Fig. 3.21 Generation of chaotic oscillations of the type [1, 0] through the first kind intermittency in the model (3.4) at $\gamma_1 = 0.7$, $\varepsilon_1 = 0$, $\gamma_2 = 0.3$, $\varepsilon_2 = 50$, and variations κ from 1.38 till 1.4.

regions of existence of chaotic oscillations is very important. As it is known (see, for example, [8–11, 60]) that at variations of the system parameters demonstrating chaotic oscillations, these chaotic oscillations can become regular and then again chaotic. In this case, regions of existence of regular oscillations within the parameter space can be isolated (surrounded by such parameter values which respond to chaotic oscillations) and rather stretched. Both these factors should

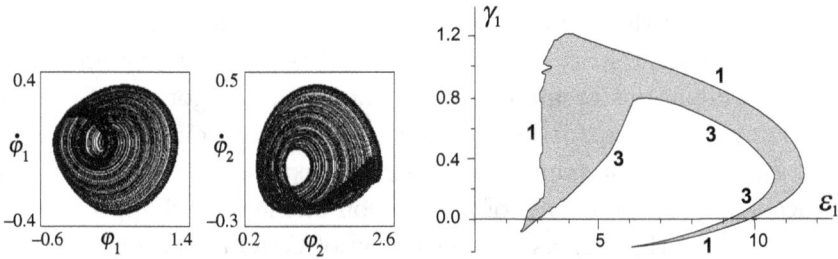

Fig. 3.22 Example of the projection of a chaotic attractor of the type [0, 0] and the region of its existence at $\gamma_2 = 0.7, \varepsilon_2 = 3, \kappa = 0.8$.

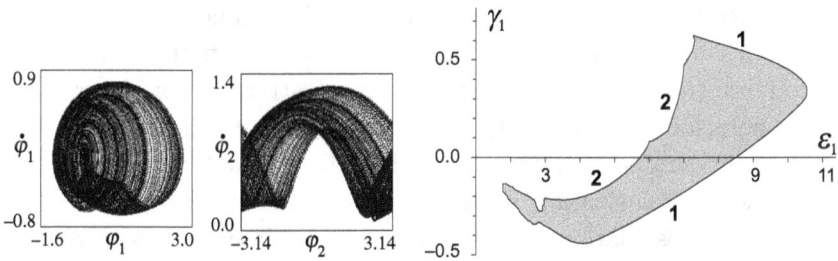

Fig. 3.23 Example of the projection of a chaotic attractor of the type [0.1] of the model (3.4) and the region of its existence at $\gamma_2 = 0.7, \varepsilon_2 = 3, \kappa = 1.6$.

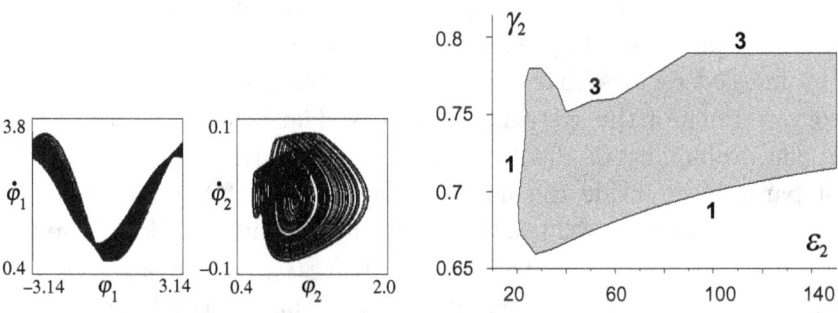

Fig. 3.24 Example of the projection of a chaotic attractor of the type [1.0] of the model (3.4) and the region of its existence at $\gamma_1 = 1.7, \varepsilon_1 = 0, \kappa = 1.9$.

be taken into account for estimation of chaotic oscillations regions. Knowledge of the internal structure of these regions allows us to not only get rid of unwanted regularization of oscillation phenomena, but also efficiently manage the properties of chaotic oscillations.

We studied the regions of existence of self-modulated oscillations in the parameter space of the model (3.4) by constructing and analyzing two-parameter maps of dynamic modes. Algorithm for constructing these maps is based on the analysis of Poincaré map and also calculating the maximum Lyapunov exponent [41, 61]. We will concentrate on the results of application of the offered approach for estimation of the regions for generation of CMO of the PLL system by the model (3.4). Figure 3.25 shows dynamic mode maps and the dominant Lyapunov exponent of the model (3.4) constructed for the analysis of regions of generation of CMO in Figs. 3.22–3.24. From the comparative analysis of Figs. 3.22–3.25, it follows that they go with each other well, however, maps of dynamic modes have additional information reflecting the internal structure of the regions for generation of CMO. Figures 3.25(a), 3.25(c) and 3.25(e) show that with parameter variations inside the marked regions, chaotic oscillation regularization is possible, however the subspace dimensions where oscillations are regular are not large. Marked regions of CMO, are, in percentage wise, as follows: in Fig. 3.25(a) (the region with the CMO mode of the type [0, 0], Fig. 3.22) 12%, and Fig. 3.25(b) (the region with the CMO mode of the type [0, 1], Fig. 3.23) 11%, in Fig. 3.25(e) (the region with the CMO mode of the type [1, 0], Fig. 3.24) 14%. Chaotic homogeneity index of the marked regions characterizes the stability of the chaotic mode to variations of the system parameters. The higher it is, the lesser is the probability of chaotic oscillation regularization at variations of parameters inside the marked regions of CMO generation. For the regions in Figs. 3.22–3.24, chaotic homogeneity indices have the following values $I_{0,0} = 0.88$, $I_{0,1} = 0.89$, $I_{1,0} = 0.86$, respectively. Maps of the dominant Lyapunov exponent characterize levels chaotic modes. From Figs. 3.25(b), 3.25(d) and 3.25(f) we can see that chaotic modes of the type [0, 0] are less chaotic in comparison with the modes of the type [0, 1] and [1, 0], as they have less Lyapunov exponent.

So, from the represented results, it follows that by the cascade coupling of two simple PLL systems (where the model attractors can be either in the equilibrium state or rotatory limit cycle), it is

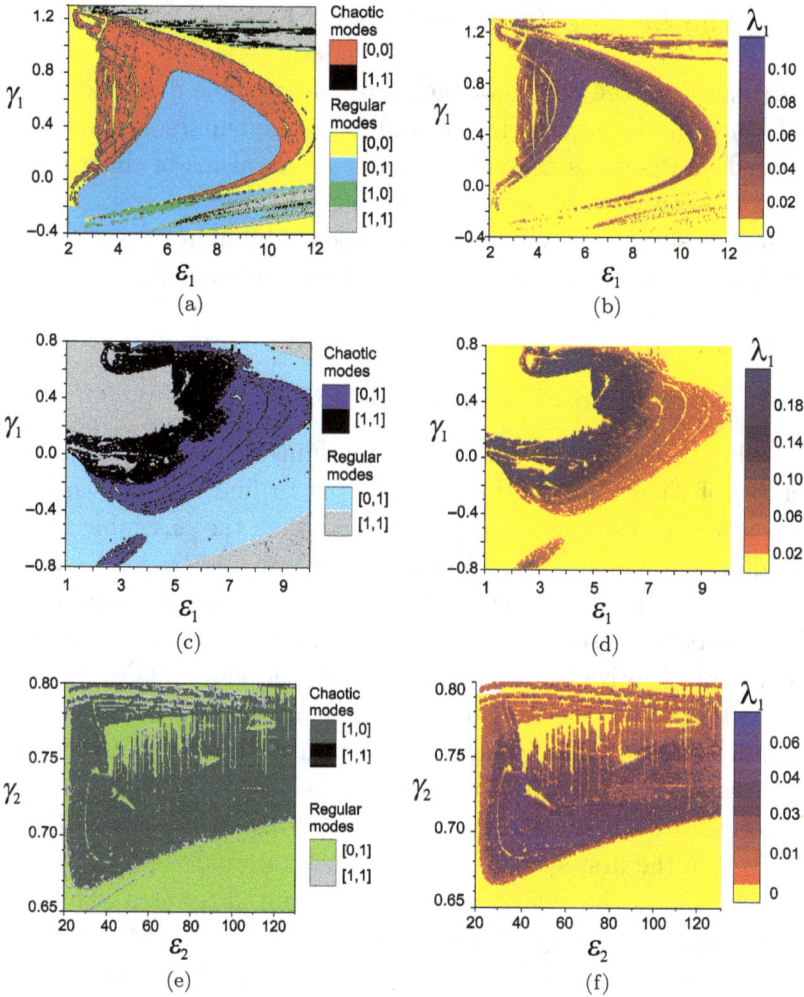

Fig. 3.25 The map of dynamical regimes and dominant Lyapunov exponent of the model (3.4) for $\gamma_2 = 0.7$, $\varepsilon_2 = 3$, $\kappa = 0.8$ (a, b); $\gamma_2 = 0.7$, $\varepsilon_2 = 3$, $\kappa = 1.6$ (c, d); $\gamma_1 = 1.7$, $\varepsilon_1 = 0$, $\kappa = 1.9$ (e, f).

possible to generate CMO at the outlet of separate generators as well as at the outlet of both generators simultaneously. It is important to notice that here the regions of CMO existence in the parameter space very much exceed the dimensions of CMO regions of one-ring PLL system.

3.6.   The Case of the Second-Order Filters

In this section, the results of the study of ensemble dynamics of two PLLs are discussed, where the individual dynamics is more complicated and permits chaotic behavior. Mechanisms of chaotic oscillation generation are examined, regions of CMO generation in the parameter space are analyzed by methods of bifurcation analysis and construction of two-parameter maps of dynamic modes.

3.6.1.   *Chaotic oscillations generation*

Modeling of the system (3.3) and examining of mechanisms of generation of chaotic oscillations allowed us to find a great variety of bifurcation transitions. It is established that the ensemble chaotic dynamic features are defined to a large extent not by a complicated dynamics of the connected systems but by couplings between the elements of the ensemble. To illustrate this fact, we will examine the process of chaotization of oscillations in the ensemble of two connected PLLs with the second-order filters at the cost of introduction of additional coupling by control circuits in the case when the coupled subsystems have a simple individual dynamics.

Let $\gamma_1 = 0.5$, $\varepsilon_1 = 1$, $\mu_1 = 1$, $\gamma_2 = 0.69$, $\varepsilon_2 = 1$, $\mu_2 = 2.37$, then at $\kappa = 0$ in the first system a synchronization mode (Fig. 3.26(a)) is realized, and in the second, depending on initial terms, there can be either a regular quasisynchronization mode (Fig. 3.26(b)), or regular mode of beatings (Fig. 3.26(c)). The one-point bifurcation diagram of Poincaré map (Fig. 3.26(a)) shows the evolution of dynamic modes at increasing κ, when the second generator functions in the quasisynchronization mode. It is clear that the introduction of additional coupling ($\kappa \approx 0.014$) leads to chaotization of oscillations at the outlet of the first generator. In this case, at the outlet of the second generator, a CMO mode is also realized (Fig. 3.26(f)). Increasing the coupling force till $\kappa = 0.07$ leads to collapse of the second generator on the chaotic mode of beatings, while the first generator continues working in the mode of generating CMO (Fig. 3.26(g)). Approaching $\kappa = 0.5$ both ensemble generators turn into modes of chaotic beatings.

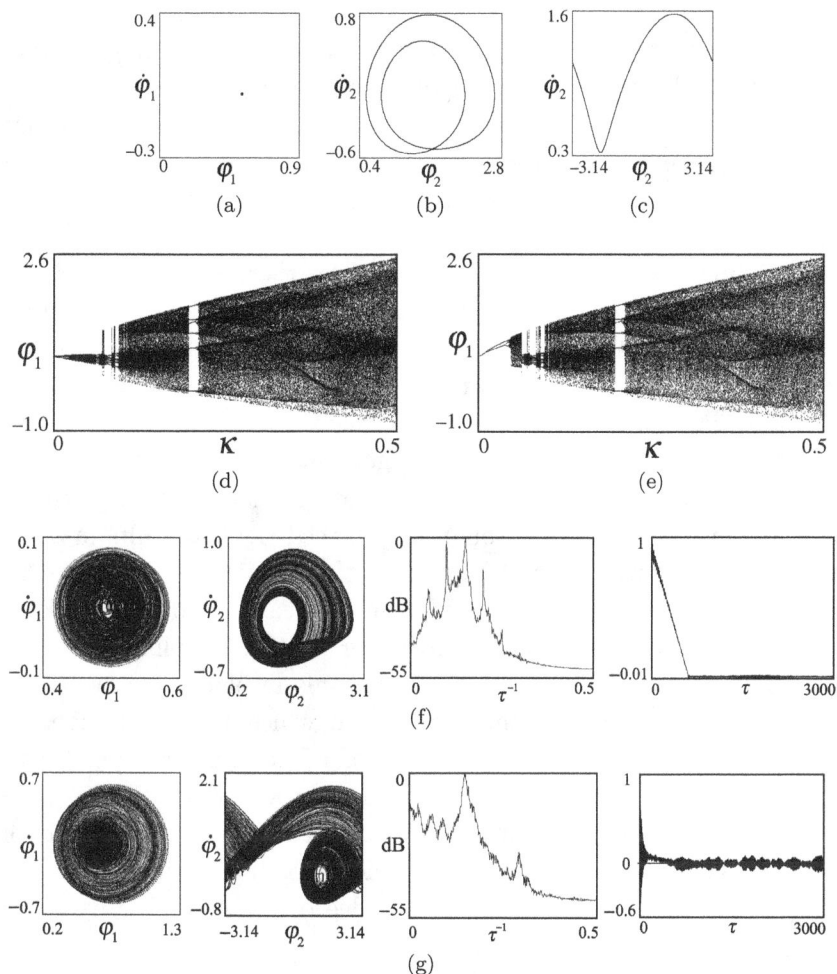

Fig. 3.26 Projections of the model (3.3) attractors at $\kappa = 0$, (a) character-izing the mode of synchronization of the first generator, (b) the mode of qua-sisynchronization and (c) the mode of beatings of the second generator; (d, e) one-parameter bifurcation diagrams of Poincaré map (3.3); projections of phase attractors portraits, spectrum and autocorrelation functions, calculated by real-ization $\varphi_1(t)$, for $\kappa = (f)$ 0.04 and (g) 0.23 respectively.

In the capacity of the initial state (at $\kappa = 0$) of the second genera-tor to choose the mode of beatings, the process of starting and devel-oping CMO at the outlet of the first generator (Fig. 3.26(e)) with increasing κ qualitatively coincides with the scenario of developing chaotic oscillations, which were examined above. Here, chaotic oscillations also start as a result of a cascade of period-doubling bifurcations at small $\kappa \approx 0.045$, where bifurcation diagram $\{\kappa, \varphi_1\}$ experiences weak changes in the interval $\kappa \in (0; 0.1)$. It is remarkable that by varying the coupling force κ, it is possible to get the second generator to function in a quasisynchronous mode, including a chaotic one.

Unifying generators into the ensemble, there is a possibility of a direct transfer from a synchronous (quasisynchronous) mode into a mode of generation of CMO. For example, in the ensemble of two PLLs let one generator function in the synchronous mode and the other in the chaotic mode. Then, realizing the coupling through unbalanced signals from the second generator to the first, it is easy to achieve generation of CMO at the outlet of the first generator. Introduction of coupling through unbalanced signals results in oscillations in the control circuit PLL_1 being modulated by chaotic oscillations of the control circuit PLL_2, in this case, modulation factor is regulated by the coupling parameter κ. If a coupling is strong enough then PLL_1 can turn on to the mode of chaotic beatings, however, independent of the scenario in which PLL_2 functions in the chaotic mode (in the CMO mode or mode of chaotic beatings), there are always such values of κ, where at the outlet of PLL_1 there will be CMO. It is remarkable that this mechanism of stimulation of chaotic oscillations is soft and is not accompanied with bifurcations of special trajectories in the phase space of the ensemble dynamic model.

So, the introduction of additional coupling allows us to generate CMO at the outlet of the first generator independently on that in which mode (quasisynchronous or beatings) the second generator functions. It is a rather important conclusion as it removes restrictions from the kind of oscillations of the second generator. Moreover, oscillatory–rotatory attractors, which are responsible for the mode of beatings of the second generator, create chaotic oscillations with a wider spectrum than the oscillatory ones at the outlet of the first generator.

3.6.2. *Chaotic oscillations regions*

We will examine the collective dynamics in the case when individual dynamics of two unified PLLs is regular (non-chaotic). We will fix the second generator parameter in the point $\gamma_2 = 0.69$, $\varepsilon_2 = 1$, $\mu_2 = 2.37$, additional coupling force on the level $\kappa = 0.1$ and examine influence the first generator parameters on the ensemble collective dynamics. As the second PLL system parameters are chosen in the point of bistable behavior, then the collective dynamics is also characterized by the bistable behavior. Here, dynamic modes can develop on the oscillatory attractors as well as on the oscillatory–rotatory ones. Figure 3.27 shows segmentations of the parameter plane (μ_1, γ_1) of the model (3.3) at $\varepsilon_1 = 1$ on the region of existence of regular and chaotic attractors, which are developed based on the oscillatory cycles (Fig. 3.27(a)) and oscillatory–rotatory cycles of the type $[0, 1]$ (Fig. 3.27(b)).

At parameter values from the region $C_0 = C_0^+ \cup C_0^-$ in the phase space of the model (3.3), there are oscillatory limit cycles which respond to the regular quasisynchronous oscillations of both ensemble generators (Fig. 3.27(a)). The region C_0^+ is situated between lines 1 and 2. Oscillatory cycles appear as a result of a saddle-node bifurcation at line 1 crossing with decreasing parameter γ_1: to the right of the point a, the cycle L_0 is created, to the left — the period-doubling cycle $L_0^{(2)}$ is created. Dotted line 3, corresponding to stability change L_0 and creation of the cycle $L_0^{(2)}$, joins the line 1 in the point a. Further decrease in the parameter γ_1 leads to the appearance of oscillatory chaotic attractor (at line 2 crossing) or as a result of series of period-doubling cycle bifurcations, or at the destruction of the oscillatory invariant torus which appears with change in stability of the cycle L_0.

In same manner, creating and chaotization of quasisynchronous oscillations occur with negative values of γ_1; at line 4 bottom-up crossing to the right of the point b the cycle L_0 is created, to the left of the point, the cycle $L_0^{(2)}$ is created. Further increase in γ_1 leads to the appearance of chaotic oscillations (at line 5 crossing)

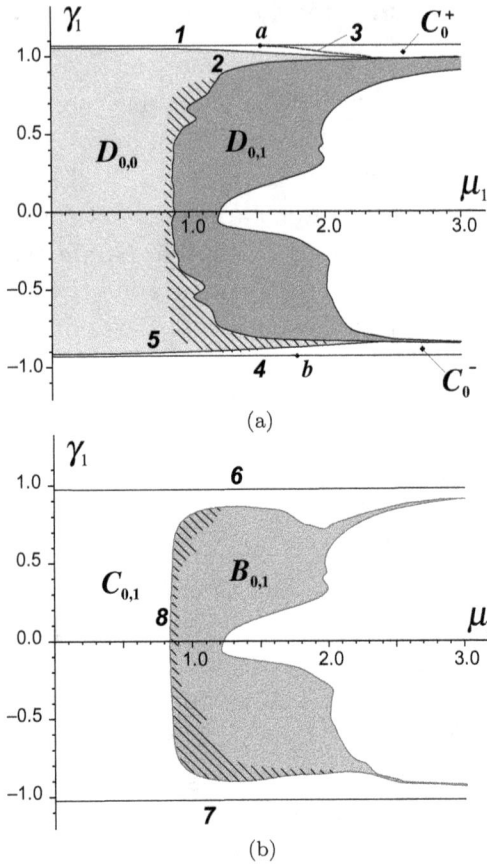

(a)

(b)

Fig. 3.27   Regions of quasisynchronous oscillations of two cascade-coupled PLLs with the second-order filters.

or according to Feigenbaum scenario or as a result of destroying the oscillatory invariant torus. In the region $D_{0,0}$, marked in Fig. 3.27(a) by light gray color, there are oscillatory chaotic attractors corresponding to CMO at the outlet of the first and the second generators of the ensemble simultaneously. Leaving the region $D_{0,0}$ through the boundary, there is bifurcation of chaotic oscillatory attractor, which turns into a chaotic oscillatory–rotatory attractor of the type $[0, 1]$, that is, rotation goes by the coordinate φ_2. This chaotic attractor exists at parameter values from the region $D_{0,1}$, marked in Fig. 3.27

by dark gray color, and corresponds to the generation of CMO at the outlet of the first generator.

At parameters from the region $C_{0,1}$ (Fig. 3.27(b)) in the phase space of the model (3.3), there is a oscillatory–rotatory limit cycle $L_{0,1}$, which corresponds to regular quasisynchronization mode of the first generator and mode of beatings of the second generator. The cycle $L_{0,1}$ appears as a result of saddle-node bifurcations at line 6 crossing with decreasing γ_1 and line 7 with increasing γ_1. Further approaching line 8, the cycle $L_{0,1}$ transforms into the chaotic attractor $S_{0,1}$ of the type $[0,1]$ by the cascade cycle of period-doubling bifurcations or by the first- or third-type intermittence or according to the following scenario: change in stability of the cycle $L_{0,1}$ → the stable oscillatory–rotatory attractors either creating or destroying $T_{0,1}$ torus creation of the chaotic attractor $S_{0,1}$. Attractor $S_{0,1}$ exists at parameter values from the region $B_{0,1}$ (dark region in Fig. 3.27(b)).

In Fig. 3.27, shaded regions indicate the collective dynamics of the examined system which is characterized by the bistable chaotic behavior. These regions represent the crossing regions $D_{0,0}$ (Fig. 3.27(a)) and $B_{0,1}$ (Fig. 3.27(b)). With parameters from this region in the CMO mode both generators as well as only the first generator can function.

From the presented results, we can see that by unifying two PLLs with a regular individual dynamics, it is possible to get a mode of CMO generation in a rather large parameter region in the first and the second generators. We will remark that in this case the obtained region for CMO generation is much larger than regions for CMO generation with individual PLL.

Now, we will examine the collective dynamics in the case when individual dynamics of the first PLL can be only regular and the second PLL has a chaotic dynamics. For this, in the control circuit of the first generator, we will choose the first-order filter and in the control circuit of the second generator we will choose the second-order filter with parameters in the region of chaotic beatings. For this case, introduction of additional coupling $\kappa \neq 0$ allows us to generate CMO at the outlet of the first generator in a wide parameter

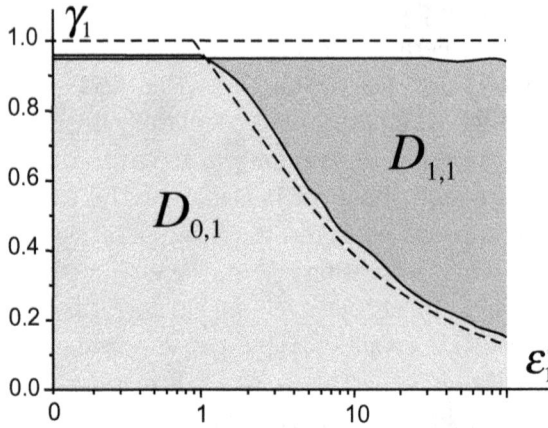

Fig. 3.28   The pull-in and pull-out ranges of CMO at the outlet of the first generator in the ensemble of two cascade-connected PLLs at $\mu_1 = 0, \gamma_2 = 0, \varepsilon_2 = 1, \mu_2 = 2.37, \kappa = 0.1$.

range. For example, in Fig. 3.28, there are chaotic mode regions of the ensemble of connected PLL for $\kappa = 0.1$. In this figure the region $D_{0,1}$ is marked; here, generation of CMO at the outlet of the first generator starts at any initial terms, and in the region $D_{1,1}$, a bistable chaotic mode is realized. At $D_{1,1}$ parameters, depending on initial terms at the outlet of the first generator, a CMO mode or mode of chaotic beatings can exist. In the examined case, the second generator always functions in the mode of chaotic beatings. Existence of regions of bistability leads to the appearance of hysteresis processes of installation and losing of the generation mode of CMO when changing the parameters of the first PLL [43]. For illustration of this phenomenon in Fig. 3.29, there is a one-parameter bifurcation diagram $\{\gamma_1, \dot{\varphi}_1\}$ of Poincaré map which proves that when changing γ_1 from 0.55 to 1 at the outlet of the first PLL, at first, there are CMO (Fig. 3.29(b)), then at $\gamma_1 = 0.96$ there is a cross from CMO to the mode of chaotic beatings (Fig. 3.29(c)), which holds at decreasing γ_1 (upper part of Fig. 3.29(a)) till $\gamma_1 = 0.58$, after that the CMO mode is installed again.

It is necessary to pay attention to the fact that the boundaries of the regions $D_{0,1}$ and $D_{1,1}$ almost coincide with the boundaries of the

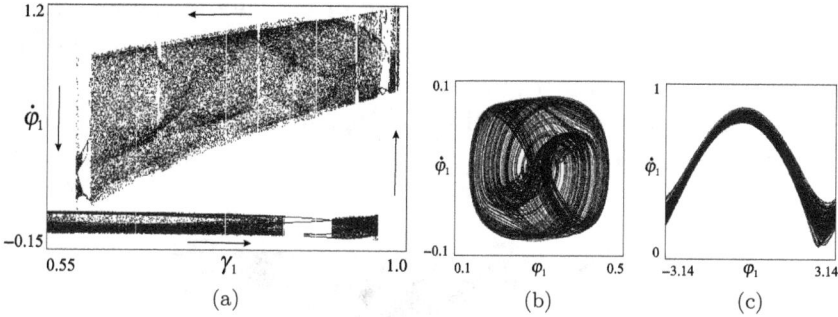

Fig. 3.29 Phenomena of hysteresis at generation CMO at the outlet of the first generator in the ensemble of two cascade-connected PLLs at $\varepsilon_1 = 5$, $\mu_1 = 0$, $\gamma_2 = 0$, $\varepsilon_2 = 1$, $\mu_2 = 2.37$, $\kappa = 0.1$.

pull-in to the regular synchronization mode region and the region of keeping the synchronization mode of individual PLL (non-chaotic), well known in the PLL theory. For comparison, in Fig. 3.28, there are boundaries of pull-in and pull-out regions of individual PLL which are marked by the dotted lines. A very important conclusion can be done based on this result: passing from the isolated PLL to the ensemble of two cascade-connected PLLs, we will surely obtain PLL generation of chaotically modulated oscillations for any parameter values, belonging to the pull-in to the synchronization mode of the individual PLL region.

3.6.3. *Self-modulated oscillations*

We refer to the analysis of the regions of existence of self-modulated modes. Analysis will be performed by examining two-parameter maps of dynamic modes of the model (3.3). In this case, main attention will be paid to the definition and analysis of the parameter regions where there are CMO (regions D_u^c of pull-out to the CMO mode), and where they are surely realized (regions D_z^c of pull-in to the CMO mode). For the estimation of the regions D_u^c and D_z^c, we will use algorithms similar to that used for calculating the pull-in and pull-out range of the PLL system into the synchronous mode, the difference being that instead of a synchronous mode a quasisynchronous mode will be analyzed.

Then, we will analyze the regions of self-modulated modes of the ensemble of two PLLs with the second-order filters and the ensemble consisting of PLL with the first- and second-order filters.

The ensemble of two PLLs with the second-order filters. In Fig. 3.30, there are maps of dynamic modes of the model (3.3)

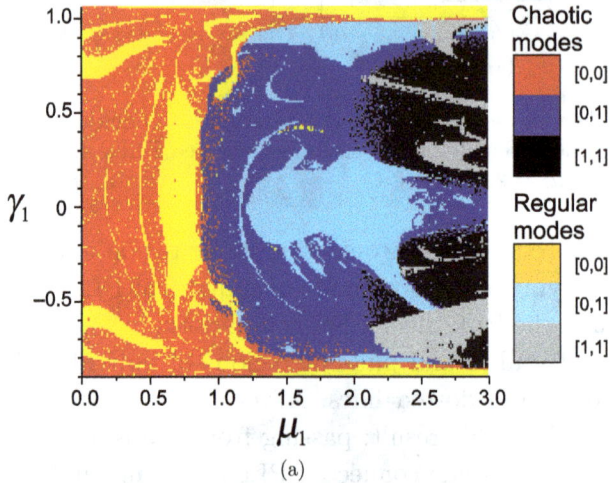

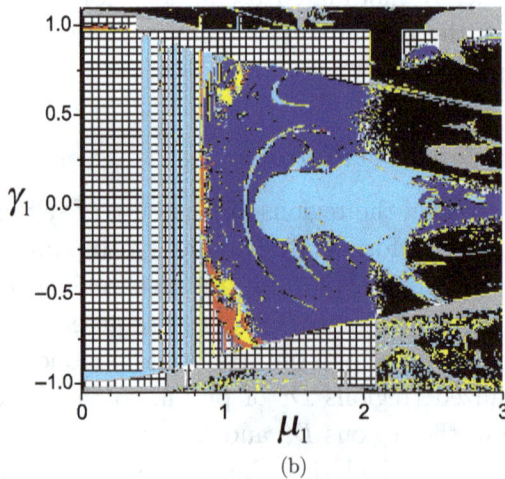

Fig. 3.30   Maps of dynamic modes of the model (3.3), which were built for estimation of the regions D_u^c and D_z^c for $\varepsilon_1 = 1$, $\gamma_2 = 0.69$, $\varepsilon_2 = 1$, $\mu_2 = 2.37$, $\kappa = 0.1$.

on the plane $\{\mu_1, \gamma_1\}$, which were built for estimation of the regions D_u^c and D_z^c. These maps are composed of $N = 64800$ points (300 horizontal points and 216 vertical points). These points differ in color which characterizes the type of the ensemble dynamic mode. The start state for constructing a map (Fig. 3.30) was the state of the model (3.3) at $\gamma_1 = 0$, $\mu_1 = 0.01$ with initial terms in the attractor of the type [0, 0]. Then, the initial frequency mismatch increased with the step $\Delta_\gamma = 0.01$ till a global mode of beatings was installed in the system; in this case, phase variable values for the next parameter value γ_1 were taken from the attractor on the previous step. Then, the described procedure was repeated for $\mu_1 \in [0, 01; 3.0]$ with the step $\Delta_\mu = 0.01$. Figure 3.30(b) is a result of stitching of two maps of dynamic modes: one is built when γ_1 changes from 1.1 to -1.0, the other when γ_1 changes from -1.05 to 1.1. In both cases, the start state of the model (3.3) was an attractor of the type [1, 1], phase variable values for the next value of the parameter γ were taken from the attractor on the previous step, parameter μ_1 varied in the interval from 0.01 to 3.0 with the step $\delta_\mu = 0.01$. Stitching of the maps allowed to single out the regions with bistable behavior (in Fig. 3.30(b) the shaded regions show bistable behavior) and regions of mono-modes.

From Fig. 3.30, it is clear that at the examined parameters, the model (3.3) has six dynamic mode types.

- Regular or chaotic mode of the type [0, 0]. PLL_1 and PLL_2 function in regular or chaotic quasisynchronous modes.
- Regular or chaotic mode of the type [0, 1]. PLL_1 functions in regular or chaotic quasisynchronous modes, and PLL_2 functions in regular or chaotic modes of beatings.
- Regular or chaotic mode of the type [1, 1]. PLL_1 and PLL_2 function in regular or chaotic modes of beatings.

From Fig. 3.30(a), it is possible to calculate the existence of the CMO mode region. For PLL_1, this region is $D_{u1}^c = (N_{[0,0]} + N_{[0,1]})/N = 0.56$, and for PLL_2 this region is $D_{u2}^c = (N_{[0,0]})/N = 0.27$. Here, $N_{[0,0]}$ and $N_{[0,1]}$ are the number of points on the dynamical mode (Fig. 3.30), which correspond to the installation of the chaotic mode types $[0, 0]$ and $[0, 1]$, N is general number of points on the map. D_{u1}^c and D_{u2}^c are the regions of the pull-out CMO mode for PLL_1 and PLL_2, respectively. The existence of regions shows that these D_{z1}^c and D_{z2}^c of the CMO mode in Fig. 3.30(b) are the regions of pull-in to the CMO mode for PLL_1 and PLL_2, respectively. The calculations show that $D_{z1}^c = 0.21$ for PLL_1 and $D_{z2}^c = 0.01$ for PLL_2.

From Fig. 3.30(a), it is clear that parameter values, which correspond to the CMO mode, are distributed in an extremely non-uniform manner. It is evident that in this case to single out the parameter regions where only chaotic oscillations are realized is very difficult. In practice, to single existence of out regions of chaos, range boundaries of parameters are often used, where with in phase spaces of mathematical model, chaotic attractors either start or die; in particular cases, they undergo internal bifurcation, that is, only an external contour of the region of a chaotic attractor existence is singled out. The regions D_u^* where chaotic oscillations exist are singled out this way. The singling out of the existence of regions of chaotic oscillations found by an external contour (Figs. 3.27 and 3.28) will be marked by an index "star"; as a rule, this will contain subregions of regular oscillations, dimensions of which can be comparable and even exceed real dimensions of the chaotic oscillation regions. This undoubtedly influences chaotic properties of the region D_u^*. Further, based on the property characteristics of D_u^* we will use the indicator of chaotic homogeneity $I = S_c/S_u$, where S_c is cumulative dimension of the region of chaotic oscillations and S_u is dimension of the region D_u^* [41].

We will estimate the indicator of chaotic homogeneity of the CMO regions D_{u1}^* and D_{u2}^* of PLL_1 and PLL_2 systems of the examined ensemble in Fig. 3.30(a). The contour limiting the regions of chaotic modes of the type $[0, 0]$ and $[0, 1]$ in Fig. 3.30(a) is rather clear and agrees well with the boundaries of the regions D_{u1}^* and D_{u2}^*, which were obtained in Sec. 3.6.2 through the numerical bifurcation analysis of the model (3.3). Dimensions of the regions D_{u1}^* and D_{u2}^* will be estimated by the number of the map points which occur in these regions. In Fig. 3.30(a), the region $D_{0,0}$ consists of $N_{0,0} = 20521$ points, corresponding to the oscillations of the type $[0, 0]$, from which $N_{0,0}^* = 15388$ points correspond to chaotic oscillations; the region $D_{0,1}$ includes $N_{0,1} = 16778$ points of the type $[0, 1]$, from which $N_{0,1}^* = 13926$ points correspond to chaotic oscillations, and $N_{1,1} = 98$ points of the type $[1, 1]$. So, the regions $D_{0,0}$ and $D_{0,1}$ contain $N_{c1} = 29314$ points, corresponding to the mode of generation of CMO of the first PLL and $N_{c2} = 15388$ points, corresponding to the mode of generation of CMO of the second PLL. It follows from here that the regions D_{u1}^* and D_{u2}^* accordingly have the following chaotic indicators $I_1^2 = 29314/(20521 + 16778) = 0.78$ and $I_2^2 = 15388/20521 = 0.68$. For comparison, we will remind that for the model of a single PLL with the second-order filter at $\varepsilon_1 = 1$ chaotic homogeneity indicator is $I = 0.73$. It is remarkable that individual dynamics of PLL_2 at examined parameters can only be regular and dimensions of pull-in and pull-out regions of CMO of a single PLL with the second-order filter at $\varepsilon_1 = 1$, calculated by the similar methods, are equal to $D_u^c = 0.065$ and $D_z^c = 0.036$, that is, extremely less than in the ensemble. Analysis of the maximum Lyapunov exponent λ_1 in Fig. 3.30(a) shows that the maximum value λ_1 accepts the chaotic movement of the type $[0, 1]$ at such values of parameters, when the individual dynamics of the united PLL is characterized by a regular quasisynchronous mode. The value of the maximum Lyapunov exponent calculated by the trajectories of the model (3.3) reaches $\lambda_1 = 0.3$, that is, three times more than possible values of the maximum Lyapunov exponent for the trajectories of the model of a partial PLL with the second-order filter.

The ensemble of two PLLs with the first- and second-order filters. We will examine the dynamics of the ensemble of two PLLs in the case when individual dynamics of PLL_1 can be regular and PLL_2 has a chaotic dynamics. For this purpose, in the control circuit of the first generator, we will choose the first-order filter and in the control circuit of the second generator we will choose the second-order filter. If we want to fix parameters of PLL_2 in the point where its individual dynamics is characterized by the mode of chaotic beatings, then with the introduction of additional coupling $\kappa_1 \neq 0$ it is possible to get generation of CMO at the outlet of the first generator in the wide parameter region. To illustrate this fact Figs. 3.31(a) and 3.31(b) show the maps of dynamic modes on

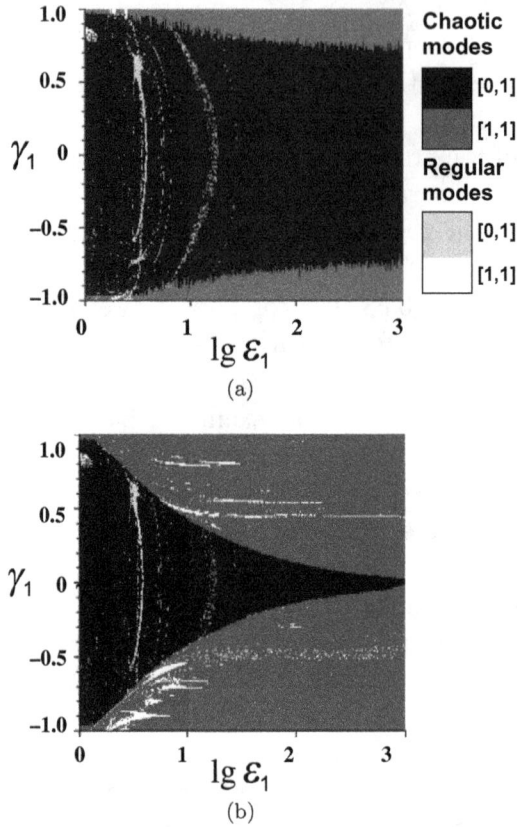

Fig. 3.31   Maps of dynamic modes of the model (3.3) at parameter values $\mu_1 = 0$, $\gamma_2 = 0.25, \varepsilon_2 = 1, \mu_2 = 6, \kappa_1 = 0.1$.

the plane $\{\varepsilon_1, \gamma_1\}$ for the ensemble which is described by Eq. (3.3) at $\mu_1 = 0$. The map in Fig. 3.31(a) was obtained by increasing the initial frequency detuning γ_1 from zero values, when there are CMO at the PLL_1 outlet, to values $\gamma_1 = \pm 1$, when a global mode of beatings is realized in the ensemble. The map in Fig. 3.31(b) was built by decreasing γ_1 from values $\gamma_1 = \pm 1$ to $\gamma_1 = 0$.

From Fig. 3.31, it is clear that the main dynamic modes of the examined ensemble are chaotic oscillations. Figure 3.31a shows the region of existence of chaotic modes of the type [0, 1] and [1, 1] which occupy, respectively, 81% and 17% of the general area of the figure. It follows that cumulative dimensions of the regions of pull-out CMO of the PLL_1 and PLL_2 systems are, respectively, equal to $D^c_{u1} = 0.81$ and $D^c_{u2} = 0$. The pull-in regions of the CMO mode, calculated by Fig. 3.31(b), are equal to $D^c_{z1} = 0.34$ and $D^c_{z2} = 0$, respectively. Chaotic indicator of the region D^*_{u1} is $I^2_1 = 0.98$. It is necessary to pay attention to the fact that the regions D^c_{z1} and D^*_{u1} practically coincide with the pull-out region of the synchronization mode of the individual PLL with the integrating filter in the control circuit which is well known in the theory of the systems PLL.

Chaotic attractors in the phase space U define modulating oscillations appearing in the control circuits of the PLL systems. They are characterized by averaged values of phase variables $\tilde{\varphi}_i$ and $\tilde{y}_i$, and also ranges of these parameter changes — rocking amplitudes $\Delta\varphi_i$ and Δy_i. Figure 3.32 shows diagrams of evolution of these modulating oscillation characteristics of PLL_1 of the ensemble of

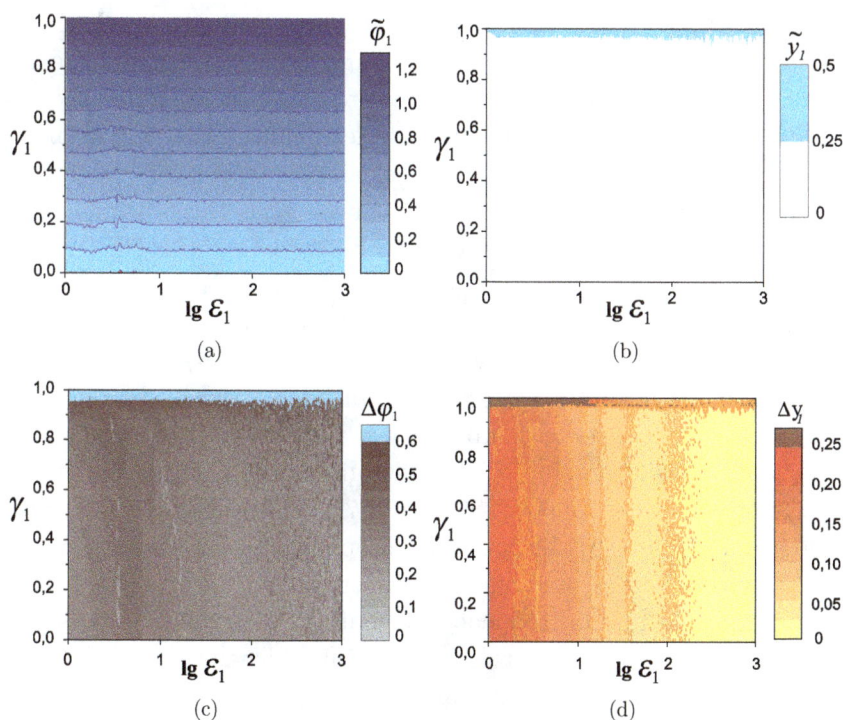

Fig. 3.32 Diagrams of evolution of modulating oscillations characteristics PLL_1 in the ensemble of two connected PLLs with the first- and second-order filters.

two connected PLLs with the first- and the second-order filters at $\gamma_2 = 0.25$, $\varepsilon_2 = 1$, $\mu_2 = 6$, $\kappa = 0.1$. The analysis of these diagrams shows that at increasing ε_1 the oscillations amplitudes $\Delta\varphi_1$ and Δy_1 become less. At $\varepsilon_1 > 100$ there is no deviation Δy_1 practically, there are only deviation $\Delta\varphi_1$.

3.7. Experimental Chaos Investigation

This section presents the results of physical experiments with the ensemble of two cascade-connected phase systems [62,63]. It is shown that the examined system is a generator of chaotic-modulated oscillations in the wide region of the parameter space.

In the capacity of a phase system, a widely spread 74HC4046A microscheme of the PLL is meant for synchronization of signals of the meander sort. It allows the constructions of the control circuit of the PLL, and defines the frequency and pull-out band of the voltage controlled oscillators. Functional scheme of the experimental installation is shown in Fig. 3.33, where RO is a reference oscillator, PD is a phase discriminator, which realizes the operation of Boolean algebra XOR, $VCO_{1,2}$ are relaxation oscillator generators whose frequencies are controlled by voltage, k is the coupling coefficient, $K_{1,2,3,4,5}(p)$ are low frequency filters with the coefficients of transmission $K_i(p) = (1 + T_i p)^{-1}, i = 1, \ldots, 5$. In the time constant experiment of the filters, $T_i = R_i C_i$ had the following values $T_1 = T_3 = 14.9$ mks ($R_1 = R_3 = 2.2$ kOhm, $C_1 = C_3 = 6.8$ nF), $T_4 = T_5 = 165$ mks ($R_4 = R_5 = 5$ kOhm, $C_4 = C_5 = 33$ nF), changing parameter $T_2 = 0–330$ mks ($R_2 = 0–10$ kOhm, $C_2 = 33$ nF). Frequency oscillations in $VCO_{1,2}$ with no external signal and noise at the outlet are as follows: $f_1 = f_2 = 71$ kHz — central frequency. The signal coming to the input of the phase system and generated on its outlet is presented in a unidirectional meander with amplitude of 5V. Maximum frequency mismatches of the input signal for $PLL_{1,2}$ concerning the central frequency at which generators $VCO_{1,2}$ can be synchronized (pull-out band $\Delta F_{1,2}$) have the following values: $\Delta F_1 = 13.85$ kHz, $\Delta F_2 = 27.7$ kHz. During the experimental measurements the following values varied: feedback gain k, directed from

(a)

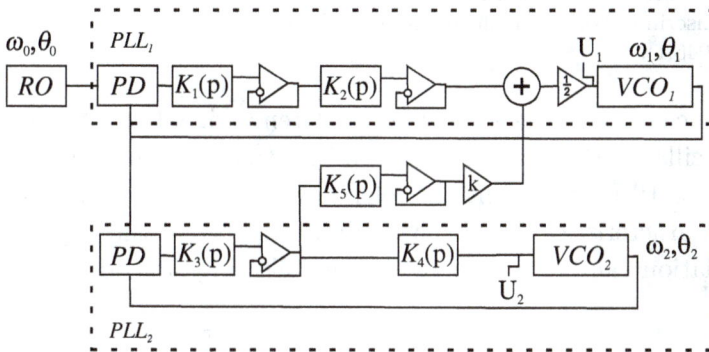

(b)

Fig. 3.33 (a) Overview and (b) functional scheme of experimental installation of two cascade-connected PLLs.

the control circuit PLL_2 into the control circuit PLL_1 ($k = 0\text{--}1$), the reference oscillator frequency f_0 and a parameter T_2 of the second filter in the control circuit PLL_1.

Figure 3.34(a) shows a diagram of static dependence of voltage Vcc at the outlet of the control circuit on difference of phases between the reference signal and the oscillator signal in PLL. The represented diagram can be interpreted as a characteristics of the phase discriminator, though a real phase comparator represents an operation of Boolean algebra XOR. Voltage oscillogram at the outlet of the comparator, on the input of which signals are generated in Figs. 3.34(b) and 3.34(c), is represented in Fig. 3.34(d). Figure 3.34(e) shows a voltage oscillogram at the outlet of a low frequency filter (and on the input VCO). It is remarkable that in the synchronous mode, the

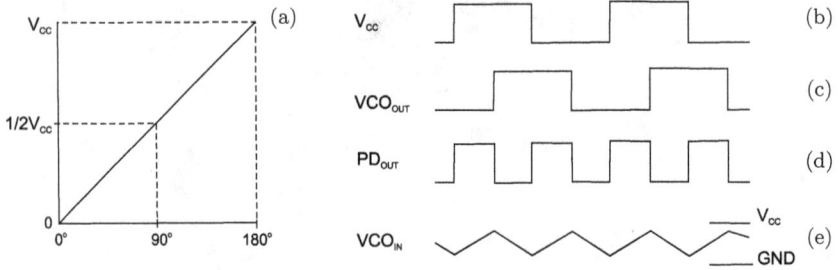

Fig. 3.34   Characteristics of (a) the phase discriminator and (b) voltage oscillogram at the input of the PLL, (c) at the outlet VCO, (d) at the outlet of the phase discriminator, (e) at the input of VCO (see description of the micro scheme 74HC4046A).

difference of the phases between a reference signal and the signal of the oscillator PLL is $\pi/2$, if the initial frequency mismatch of the oscillator PLL concerning the reference signal is equal to 0. Normalized characteristics of the phase discriminator of the experimental installation can be represented in the following dependence:

$$F_{\mathrm{XOR}}(\varphi) = \begin{cases} 2\varphi/\pi - 1, & 0 < \varphi \leq \pi; \\ 3 - 2\varphi/\pi, & \pi < \varphi \leq 2\pi. \end{cases}$$

To study different dynamic modes, two digital phase detectors were constructed to define the difference between the reference signal and the signal at the outlet of the oscillators in $\mathrm{PLL}_{1,2}$. It allowed to track phase shifts and divide self-modulated modes into quasisynchronous and asynchronous.

In the presented experimental scheme, two cascade-connected PLLs have filters of the type $[0/2]$, realized in the capacity of two serially connected RC filters. These filter parameters are chosen such that the PLL systems have extremely regular individual dynamics, for example, in the first PLL at the examined parameters either a synchronous or regular mode of beatings is realized. Figure 3.35 shows experimental and theoretical curves which are the boundaries of the pull-in region in the PLL_1 system. Dotted lines are used to draw the boundaries which were received from the results of computer modeling and full lines are the result of the real experiment. Here, lines 1 and 2 are calculated by the model (2.2) at $n = 0$,

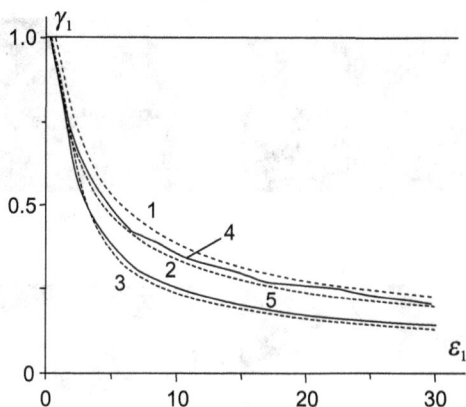

Fig. 3.35   Theoretical and experimental dependencies of the pull-in range of the PLL system.

$F(\varphi) = \sin \varphi$ and $F(\varphi) = F_{\text{XOR}}(\varphi)$, respectively; line 3 is by the model (2.3) at $n_1 = n_2 = 0, F(\varphi) = F_{\text{XOR}}(\varphi)$; lines 4 and 5 are received for PLL_1 with the integrating filter ($T_1 = 0$) and for PLL_2 with the second-order filter. As it can be seen from Fig. 3.35, the model of the continuous PLL system can be used to describe this experimental installation. While using the function of the phase discriminator, $F(\varphi) = F_{\text{XOR}}(\varphi)$, a good agreement is seen between the experiment and numerical modeling. As a result of performed experiments, it was established that in the second PLL of the ensemble with no additional couplings except the synchronous mode and the mode of beatings, a regular quasisynchronous mode is possible. In this mode, the VCO frequency of the second system is changed periodically concerning some medium, which is equal to the reference signal frequency, in this case, the difference in the phases $\theta_2 - \theta_0$ is limited ($|\theta_2 - \theta_0| < 2\pi$).

Figure 3.36 shows projections of the phase portraits of the system for different values of the feedback parameter k, the reference oscillation frequency $f_0 = 69.6$ kHz, $R_2 = 5$ kOhm. Along the vertical axis, there exist values of voltage on the input of the controlled oscillators $U_{1,2}$, which are proportional to momentary differences of frequency of controlled oscillators and the reference oscillations; along the horizontal axis, differences of phases $\theta_1 - \theta_0$ and $\theta_2 - \theta_0$ exist. So, the

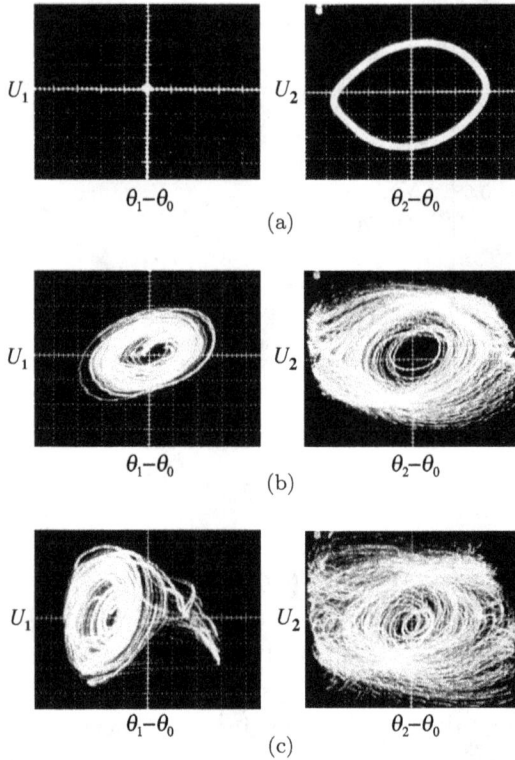

Fig. 3.36 Projections of phase portraits of the ensemble of two cascade-connected PLLs (experiment).

figures on the left correspond to the projections on the variables for the first phase system, and the ones on the right correspond to the second phase system. The phase portrait in Fig. 3.36(a) reflects the situation when there is no coupling between phase systems, PLL_1 is in the synchronization mode, and PLL_2 is in a regular quasisynchronous mode.

Introduction of the feedback leads to the appearance of oscillations in the control circuit of the first phase system, which becomes chaotic at rather little value of the coupling parameter $k \approx 0.04$ (chaotic mode diagnostics are performed visually by examining the projections of phase portraits). As the cross to chaos occurs in an extremely narrow region of the parameter k, then the scenario of the cross is not observed experimentally, we can only remark that it is soft, that is, chaotic attractor appears in the surroundings of the regular limit movement which loses stability. Figure 3.36(b) shows the projections of phase portraits for the parameter value of feedback $k = 0.3$. In the first system, there are chaotic oscillations; in this case the difference of the phase $|\theta_1 - \theta_0|$ is limited. Such mode corresponds to the generation at the outlet of the first oscillator of quasisynchronous oscillations, the frequency of which is changed chaotically around some medium one, stabilized corresponding to the reference signal frequency. In the second system, the difference of phase $|\theta_2 - \theta_0|$ is not limited; these oscillations correspond to the mode of chaotic beatings. With increase in the parameter $k \geq 0.6$, the first system also comes into the mode of chaotic beatings (Fig. 3.36(c), $k = 0.6$). In the examined experimental scheme at limits upon the parameters ($f_1 = f_2$, $k = 0 - 1$, $R_2 = 0 - 10$ kOhm and other fixed parameters), the following chaotic modes can be observed: either generation of CMO in the first phase system and generation of oscillations in the mode of chaotic beatings in the second system, or generation in the mode of chaotic beatings in both systems.

As a result of the investigation it was established that the examined scheme is able to generate align CMO at the outlet of the first phase system in the wide parameter region. The region of existence of such mode is shown in Fig. 3.37 in the plane of parameters (γ, ε_2) at $k = 0$, where $\gamma = (f_1 - f_0)/\delta F_1$, $\varepsilon_2 = R_2 C_2 \delta F_1$. The similar region of existence of such mode is for negative values γ. It is possible to remark that the obtained region for generation of CMO oscillations conforms well with the results of quality-numerical investigation.

Figure 3.38 shows the experimentally measured power spectrum of chaotic oscillations at the outlet of the first PLL system in the surroundings of the first frequency component of the reference signal which is in this case equal to $f_0 = 69$ kHz, $R_2 = 5$ kOhm. In Fig. 3.38(a), the spectrum of quasisynchronous chaotic-modulated oscillation ($k = 0, 3$) has a brightly expressed maximum corresponding to the frequency of the reference signal. In this mode of generation, it is possible to control the shift of the whole spectrum of CMOs by changing the frequency of the reference signal within the

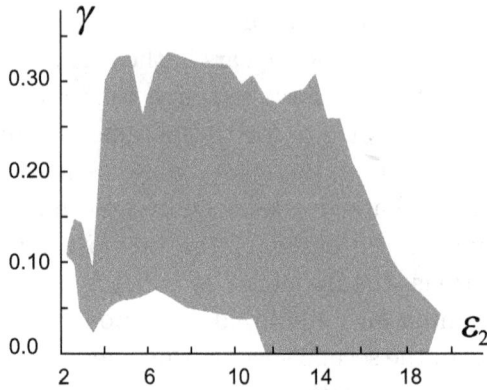

Fig. 3.37   The region of CMO generation at the outlet of the first generator of the ensemble of two cascade-connected PLLs (experiment).

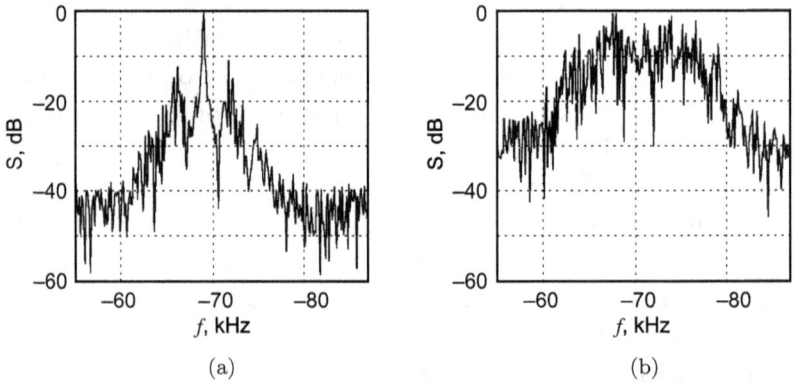

Fig. 3.38   Experimental signal spectrum: (a) chaotic modulated and (b) chaotic beatings.

limits of the region of this mode existence (Fig. 3.37). In Fig. 3.38b, the spectrum of chaotic oscillations at the outlet of the first phase system corresponds to the mode of chaotic beatings ($k = 1$). In this case, spectrum is broadened and regular; there are no bright peaks. This spectrum makes the mode of generation of chaotic beatings also interesting from the point of view of use in incoherent communication system as well as for application in radiolocation.

Chapter 4

Three Cascade-Coupled Phase System Dynamics

This chapter deals with the dynamic behavior of a three cascade-coupled phase-locked loop (CPLL) ensemble. It continues the study of nonlinear dynamics of small coupled phase-system ensembles, which was initially discussed in Chapter 3. In Chapter 4, the main focus was placed on the new phenomena caused by an increased number of coupled subsystems. We will consider ensemble models, whose dimensions do not exceed the dimensions of the previously considered examples. Particularly, the authors undertake the analysis of a three-component ensemble, in which each of the components has exclusive regular dynamics.

4.1. Mathematical Models and Dynamic Modes

The structural scheme of an ensemble formed by three cascade-coupled PLL (CPPL) systems is shown in Fig. 4.1 [53]. This scheme provides parameter tuning ($\theta_i(t)$ phase and $S_i(t)$ oscillations $\omega_i(t) \equiv d\theta_i/dt$ frequency) of G_i voltage controlled oscillators for the $S(t)$ reference signal parameters ($\theta_0(t)$ phase and $\omega_0(t) \equiv d\theta_0/dt$ frequency). The main components of this structural scheme include G_i tuned generators, which generate $S_i(t)$ oscillations, PD_i phase discriminators of $\psi_i(t) = \theta_i(t) - \theta_{i-1}(t)$ phase mismatch values, control circuits with low pass filters (F_i) and generator frequency control circuits (C_i). "Back" couplings are realized through the transmission of phase mismatching signals from PD_2 and PD_3, via some κ_1 and κ_2 converting elements, to the control circuit of the first and the second generators, respectively. In operator form, dynamic equations for this

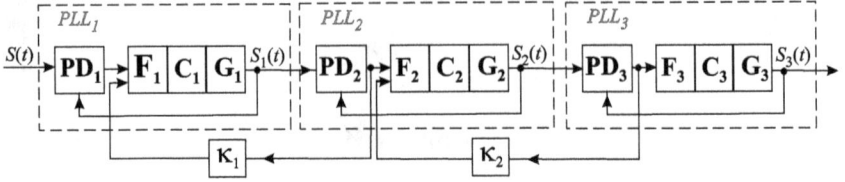

Fig. 4.1   Cascade coupling of three PLLs.

type of system at the sin-wave phase discriminator characteristics can be expressed as [4, 53]:

$$\frac{p\varphi_1}{\Omega_1} = \gamma_1 - K_1(p)[\sin\varphi_1 - \kappa_1\sin(\varphi_2 - \varphi_1)],$$

$$\frac{p\varphi_2}{\Omega_2} = \gamma_2 - K_2(p)[\sin(\varphi_2 - \varphi_1) - \kappa_2\sin(\varphi_3 - \varphi_2)], \qquad (4.1)$$

$$\frac{p\varphi_3}{\Omega_3} = \gamma_3 - K_3(p)\sin(\varphi_3 - \varphi_2),$$

where $p \equiv d/dt$, $\varphi_i = \theta_i - \theta_0$ are phase mismatch values of generators G_i for the reference signal, $\gamma_i = \Omega_i^0/\Omega_i$ are initial frequency mismatch values of the i generator for the reference signal, Ω_i are the maximum frequency mismatch values, which can compensate local control circuits, and $K_i(p)$ ($i = 1, 2, 3$) are transmission functions of the low pass filters.

Hereinafter, for the sake of simplicity, we will examine an ensemble consisting of similar elements with the simplest control circuit filters, i.e., at $\Omega_1 = \Omega_2 = \Omega_3 = \Omega$ and $K_i(p) = (1 + T_i p)^{-1}$ (where T_i are the filter time constants). In this case, we can derive the following CPLL mathematical model from (4.1):

$$\frac{d\varphi_1}{d\tau} = y_1, \quad \varepsilon_1\frac{dy_1}{d\tau} = \gamma_1 - y_1 - \sin\varphi_1 - \kappa_1\sin(\varphi_2 - \varphi_1),$$

$$\frac{d\varphi_2}{d\tau} = y_2, \quad \varepsilon_2\frac{dy_2}{d\tau} = \gamma_2 - y_2 - \sin(\varphi_2 - \varphi_1) - \kappa_2\sin(\varphi_3 - \varphi_2),$$
$$\tag{4.2}$$

$$\frac{d\varphi_3}{d\tau} = y_3, \quad \varepsilon_3\frac{dy_3}{d\tau} = \gamma_3 - y_3 - \sin(\varphi_3 - \varphi_2),$$

where $\tau = \Omega t$, $\varepsilon_i = \Omega T_i$. The system (4.2) has been determined on the six-dimensional phase cylinder $V_6 = \{\varphi_1 \,(\mathrm{mod}\,2\pi),\ y_1,\ \varphi_2 \,(\mathrm{mod}\,2\pi),$ $y_2,\ \varphi_3 \,(\mathrm{mod}\,2\pi),\ y_3\}$ in the eight-dimensional parameter space $\Lambda_1 = \{\gamma_1, \varepsilon_1, \gamma_2, \varepsilon_2,\ \gamma_3,\ \varepsilon_3, \kappa_1, \kappa_2\}$. This CPLL model is characterized by a great variety of stationary motions, which correspond to its various stable dynamic modes. Interestingly, this variety of motions is forced by the presence of three cyclic phase coordinates rather than by the model dimension (it is well known that chaotic motions can be realized even in three-dimensional models). Therefore, when considering the problem of model attractor compliance with generator dynamic modes, we will first turn our attention to the CPLL model with low-inertia control circuits.

At $\varepsilon_1 \ll 1$, $\varepsilon_2 \ll 1$, $\varepsilon_3 \ll 1$, the system (4.2) is a small parameter system at the $dy_i/d\tau$ derivatives therefore, the motions within the U phase space are subdivided into "rapid" and "slow" [38]. The surface of the "slow" motions Z is determined by the following equations:

$$y_1 = \gamma_1 - \sin\varphi_1 - \kappa_1 \sin(\varphi_2 - \varphi_1),$$
$$y_2 = \gamma_2 - \sin(\varphi_2 - \varphi_1) - \kappa_2 \sin(\varphi_3 - \varphi_2),$$
$$y_3 = \gamma_3 - \sin(\varphi_3 - \varphi_2).$$

It is stable for "rapid" motions. Equations for slow motions on the Z surface have the following expression:

$$\frac{d\varphi_1}{d\tau} = \gamma_1 - \sin\varphi_1 - \kappa_1 \sin(\varphi_2 - \varphi_1),$$
$$\frac{d\varphi_2}{d\tau} = \gamma_2 - \sin(\varphi_2 - \varphi_1) - \kappa_2 \sin(\varphi_3 - \varphi_2), \qquad (4.3)$$
$$\frac{d\varphi_3}{d\tau} = \gamma_3 - \sin(\varphi_3 - \varphi_2).$$

The system (4.3) is defined on a three-dimensional phase torus $V_3 = \{\varphi_1 \,(\mathrm{mod}\,2\pi), \varphi_2 \,(\mathrm{mod}\,2\pi), \varphi_3 \,(\mathrm{mod}\,2\pi)\}$ in a five-dimensional parameter space $\Lambda_2 = \{\gamma_1,\ \gamma_2,\ \gamma_3, \kappa_1, \kappa_2\}$.

Like collective dynamics of a two-phase-locked loop (PLL) ensemble, the collective dynamics of a three-PLL ensemble is determined by the attractors, which exist in the global phase spaces of respective mathematical models. We will characterize these dynamic modes through respective rotation indices (quasisynchronism index), which will now consist of three numbers $[J_1, J_2, J_3]$. Dynamics of each individual PLL included in the ensemble is determined by attractor projections on respective local subspaces, whose dimensions in models (4.2) and (4.3) are, respectively, equal to two and one. Interestingly, individual PLL dynamics within a three-PLL ensemble is exceptionally simple and does not allow for chaotic behavior.

The task of identifying the correspondence between mathematical model attractors and controlled generator dynamic modes is a priority objective of the study focused on the collective behavior of coupled PLLs. Further examination of the motions, which occur in the phase space following the change of the model parameters, enables the scientists to identify the basic laws of collective behavior and find out how control circuit coupling scheme and parameters can influence controlled generator dynamics. Due to sufficient nonlinearity and high dimensionality of models (4.2) and (4.3), the above-formulated task was mainly solved by numerical simulation based on the qualitative theory techniques and multidimensional dynamic systems bifurcation theory with the help of an original software system.

4.2.   Synchronous Modes

Ensemble generators can operate in the synchronization mode for the reference signal both together and separately. Consequently, we can differentiate between *global synchronization* of the ensemble — a mode, in which all controlled generators are synchronized for the reference signal — and its *partial synchronization*, which involves only individual generators. Stable equilibrium states of respective mathematical models correspond to *global* synchronization modes. In this case, equilibrium state coordinates based on φ_1, φ_2 and φ_3 variables characterize the errors of synchronization of tuned up generators for the reference signal.

Let us consider equilibrium states of the model (4.3). At the parameter values belonging to the C_0 region, the V_3 phase torus has eight equilibrium states:

$$O_1(\varphi_1^1 = q_1, \varphi_2^1 = q_1 + q_2, \varphi_3^1 = q_1 + q_2 + q_3),$$
$$O_2(\varphi_1^2 = q_1, \varphi_2^2 = \pi + q_1 - q_2, \varphi_3^2 = \pi + q_1 - q_2 + q_3),$$
$$O_3(\varphi_1^3 = \pi - q_1, \varphi_2^3 = -q_1 - q_2, \varphi_3^3 = -q_1 - q_2 + q_3),$$
$$O_4(\varphi_1^4 = \pi - q_1, \varphi_2^4 = \pi - q_1 + q_2, \varphi_3^4 = \pi - q_1 + q_2 + q_3),$$
$$O_5(\varphi_1^5 = q_1, \varphi_2^5 = q_1 + q_2, \varphi_3^5 = \pi + q_1 + q_2 - q_3),$$
$$O_6(\varphi_1^6 = q_1, \varphi_2^6 = \pi + q_1 - q_2, \varphi_3^6 = q_1 - q_2 - q_3),$$
$$O_7(\varphi_1^7 = \pi - q_1, \varphi_2^7 = -q_1 - q_2, \varphi_3^7 = \pi - q_1 - q_2 - q_3),$$
$$O_8(\varphi_1^8 = \pi - q_1, \varphi_2^8 = \pi - q_1 + q_2, \varphi_3^8 = -q_1 + q_2 - q_3),$$

where $q_1 = \arcsin[\gamma_1 - \kappa_1(\gamma_2 - \kappa_2\gamma_3)]$, $q_2 = \arcsin[\gamma_2 - \kappa_2\gamma_3]$ and $q_3 = \arcsin\gamma_3$. The equilibrium state existence region at $\gamma_3 \neq 0$ is defined by the following expressions:

$$C_0 = \{\kappa_2^- < \kappa_2 < \kappa_2^+, |\gamma_3| < 1\},$$
$$\kappa_2^- = \max\left[\frac{-1 \cdot \text{sign}\gamma_3 + \gamma_2}{\gamma_3}, \frac{-1 \cdot \text{sign}(\gamma_3\kappa_1) - \gamma_1 + \gamma_2\kappa_1}{\gamma_3\kappa_1}\right], \quad (4.4)$$
$$\kappa_2^+ = \min\left[\frac{1 \cdot \text{sign}\gamma_3 + \gamma_2}{\gamma_3}, \frac{1 \cdot \text{sign}(\gamma_3\kappa_1) - \gamma_1 + \gamma_2\kappa_1}{\gamma_3\kappa_1}\right].$$

At $\gamma_3 = 0$, the C_0 region does not depend on the κ_2 parameter and coincides with the equilibrium state existence region of a two cascade-coupled generator mathematical model [57].

The diagram in Fig. 4.2 shows the distribution of the synchronous mode existence regions (model (4.3)) on the (κ_1, κ_2) plane at $\gamma_1 = \gamma_2 = \gamma_3 = 0.5$. The dotted line outlines the C_0 equilibrium state existence region borders, h_1, h_2, h_3, h_4 lines correspond to Andronov–Hopf bifurcations and divide C_0 into C_1, C_3, C_6 and C_8 regions with stable O_1, O_3, O_6 and O_8 equilibrium states, respectively. The stable O_1, O_3, O_6 and O_8 equilibrium states correspond to the I_1, I_3, I_6 and I_8 global synchronization modes. The latter differ from each other

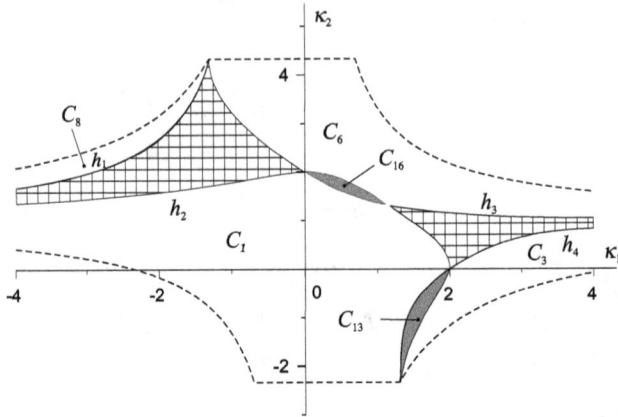

Fig. 4.2   Distribution of the global synchronization mode existence regions on the additional coupling parameter plane (κ_1, κ_2) at $\gamma_1 = \gamma_2 = \gamma_3 = 0.5$.

by individual generator synchronization errors. From the analysis of Fig. 4.2, it follows that the introduction of additional couplings can result in either the disappearance of synchronous modes (shaded areas) or the coexistence of two synchronous modes (in the $C_{16} = C_1 \cap C_6$ $C_{13} = C_1 \cap C_3$ regions).

CPLL *partial synchronization* modes can be realized in the total absence of additional couplings $\kappa_1 = \kappa_2 = 0$ or in the absence of one of the additional couplings $\kappa_1 = 0$, $\kappa_2 \neq 0$ or $\kappa_1 \neq 0$, $\kappa_2 = 0$. When $\kappa_1 = \kappa_2 = 0$ at $|\gamma_{1,2}| < 1$ and $|\gamma_3| > 1$, we can observe the first and second generators' synchronization mode, whereas at $|\gamma_1| < 1$, $|\gamma_2| > 1$ we can observe only the first generator synchronization mode. At $\kappa_1 = 0, \kappa_2 \neq 0$ the mode of the partial CPLL synchronization is conditioned by the synchronization of the first generator. This mode is realized for $|\gamma_1| < 1$. At $\kappa_1 \neq 0, \kappa_2 = 0$, the first and the second generators can be tuned up to the reference signal simultaneously. The conditions required for this type of synchronization coincide with those of the global synchronization of a two-coupled generator ensemble (see Sec. 2.4.1). Partial CPLL phase space synchronization modes are connected either with limit cycles or stable invariant varieties, which can be both regular and chaotic. Attractors responsible for the partial CPLL synchronization modes are

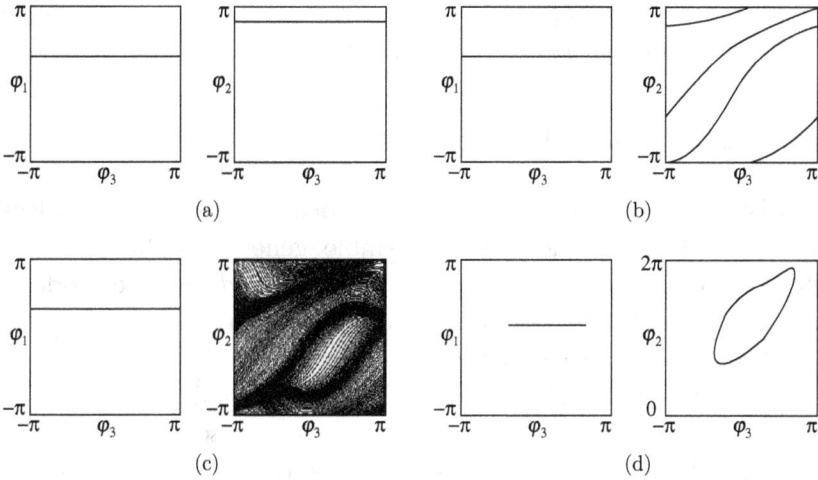

Fig. 4.3 Projections of the model attractors (4.3), which correspond to CPLL partial synchronization modes at (a) $\gamma_1 = 0.9$, $\gamma_2 = 0.99$, $\gamma_3 = 1.5$, $\kappa_1 = 0.5$, $\kappa_2 = 0.5$; (b) $\gamma_1 = 0.9$, $\gamma_2 = 1.5$, $\gamma_3 = 1.5$, $\kappa_1 = 0$, $\kappa_2 = 0.5$; (c) $\gamma_1 = 0.8$, $\gamma_2 = 0.99$, $\gamma_3 = 1.5$, $\kappa_1 = 0$, $\kappa_2 = 1.1$; (d) $\gamma_1 = \gamma_2 = \gamma_3 = 0.5$, $\kappa_1 = 0$, $\kappa_2 = 1.8$.

characterized by the fact that at least one of their two-dimensional projections on the (φ_i, φ_j) plane is a straight line segment with the length of $l \leq 2\pi$, parallel to the abscissa or the ordinate axis, where i, j take up values ranging from 1 to 3. Figure 4.3 shows sample projections of regular attractors responsible for partial CPLL synchronization. Figure 4.3(a) illustrates the establishment of a synchronous mode in the first and second generators, while Figs. 4.3(b)–(d) show the establishment of this mode only in the first generator.

Analysis of global CPLL synchronization modes with inertia control circuits carried out for the model equilibrium states (4.2) shows that even an insignificant increase in the $\varepsilon_1, \varepsilon_2$ and ε_3 inertia parameters causes significant reduction of the global CPLL synchronization mode region. Regions C_6, C_8 and C_3 undergo the most significant transformation. These changes result in the extinction of bistable synchronous behavior regions C_{16} and C_{13}. Analysis of the model attractors (4.2), which correspond to the partial CPLL synchronization modes, shows that the increase in the PLL control circuit T inertias can initiate new modes. One of the possible modes is the mode in which one of the generators operates in the synchronous mode while two others generate chaotic oscillations.

4.3. Regular Quasisynchronous Modes

Quasisynchronous, as well as synchronous, modes can be realized either in all or in individual ensemble generators. In the phase spaces, the CPLL *global quasisynchronous mode* — the mode, in which all generators operate in a quasisynchronous mode — is matched by oscillatory attractors (limited for the φ_1, φ_2 and φ_3 phase coordinates). The latter can be represented either by oscillatory limit cycles of varying complexity or by oscillatory invariant tori, including multidimensional tori. The escape of the phase trajectory onto the oscillatory limit cycle is equivalent to the generation of single-frequency modulations by all the ensemble subsystems (single-frequency quasisynchronous mode). The phase trajectory of the two-dimensional oscillatory torus establishes a two-frequency modulation mode (two-frequency quasisynchronous mode) at the output of all ensemble generators; the phase trajectory of the three-dimensional oscillatory torus establishes a three-frequency modulation mode (three-frequency quasisynchronous mode), etc. Figure 4.4 shows attractors, which correspond to global quasisynchronization modes.

It has been established that in CPLL models, stable oscillatory cycles appear when equilibrium states undergo a soft loss of stability (Andronov–Hopf bifurcation) resulting from saddle-node bifurcation and bifurcations of separatrix loops, which do not encircle the V_6 (torus V_3) phase cylinder itself.

Partial quasisynchronization modes are the modes in which the ensemble includes generator functioning both in quasisynchronous and in beat modes for the reference signal. In dynamic system phase spaces, such modes are connected with oscillatory–rotatory attractors (limit cycles of varying complexity, two-dimensional and multi-dimensional tori).

Figure 4.5 shows projections of the attractors responsible for the establishment of various regular quasisynchronous modes

(a)

(b)

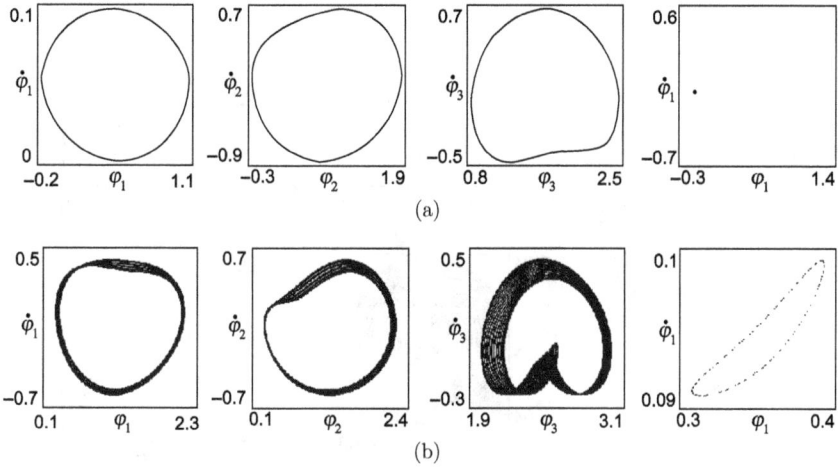

Fig. 4.4 Projections of CPLL model attractors and Poincaré maps corresponding to global quasisynchronous modes: (a) single-frequency at $\gamma_1 = \gamma_2 = \gamma_3 = 0.5$, $\kappa_1 = 0.5$, $\kappa_2 = 0.63$, $\varepsilon_1 = \varepsilon_2 = \varepsilon_3 = 1$ and (b) two-frequency $\gamma_1 = \gamma_2 = 0.7$, $\gamma_3 = 0.8$, $\kappa_1 = 1.1$, $\kappa_2 = 0.98$, $\varepsilon_1 = 0.97$, $\varepsilon_2 = 0.08$, $\varepsilon_3 = 0.1$.

in individual ensemble generators. The limit cycle presented in Fig. 4.5(a) has the rotation index equal to $[0, 1, 1]$. It can be observed at $\gamma_1 = \gamma_2 = \gamma_3 = 0.5$, $\kappa_1 = 0.1$, $\kappa_2 = 3.1$, $\varepsilon_1 = \varepsilon_2 = \varepsilon_3 = 0$ and corresponds to the establishment of single-phase oscillations: quasisynchronous oscillations in the first generator and beats in the second and third generators. The two-dimensional torus with the $[1, 0, 1]$ rotation index presented in Fig. 4.5(b) exists at $\gamma_1 = 0.996$, $\gamma_2 = 0.21$, $\gamma_3 = 0.5$, $\kappa_1 = 1.4$, $\kappa_2 = 0.35$, $\varepsilon_1 = 0, \varepsilon_2 = 50, \varepsilon_3 = 65$ and determines two-frequency modes: a quasisynchronous mode in the second generator and a beat mode in the first and third generators. The model attractor (4.2) with the $[1, 0, 0]$ rotation index, plotted at $\gamma_1 = 1.7, \gamma_2 = 0.3, \gamma_3 = 0.5$, $\kappa_1 = 1.4$, $\kappa_2 = 0.35$, $\varepsilon_1 = 0, \varepsilon_2 = 50, \varepsilon_3 = 65$ (Fig. 4.5(c)), is responsible for the establishment of the two-frequency beat mode in the first generator, two-frequency quasisynchronous mode in the second generator and single-frequency quasisynchronous mode in the third generator. The attractors (4.2), whose phase portraits are presented in Fig. 4.5(d) for $\gamma_1 = 0.14, \gamma_2 = -1.1$, $\gamma_3 = 0.07$, $\kappa_1 = 0.85$, $\kappa_2 = 3.9$, $\varepsilon_1 = 20$ and $\varepsilon_2 = 6, \varepsilon_3 = 65$ parameter values, and in Fig. 4.5(e) for $\gamma_1 = 0.6, \gamma_2 = 1.8$, $\gamma_3 = 1.4$, $\kappa_1 = 0.5$, $\kappa_2 = 0.8$,

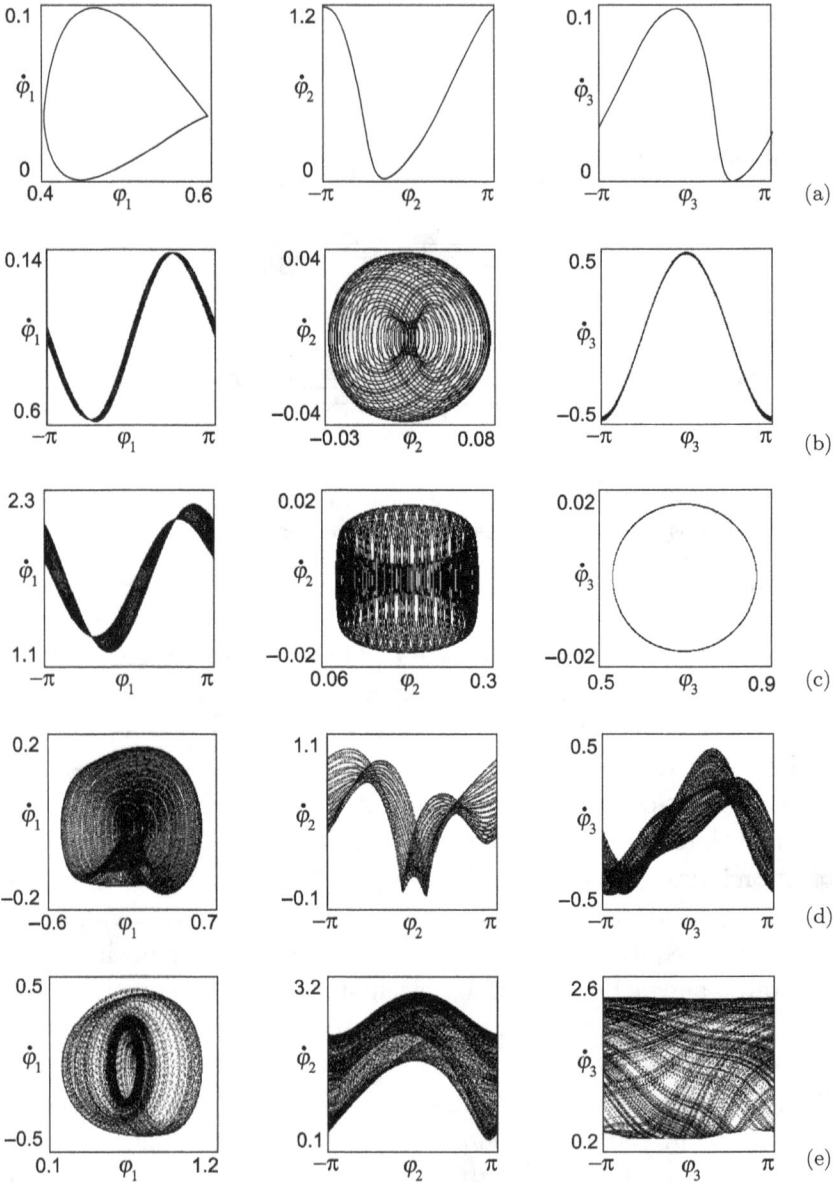

Fig. 4.5   Examples of attractors corresponding to partial CPLL quasisynchronization modes.

(a)

(b)

(c)

(d)

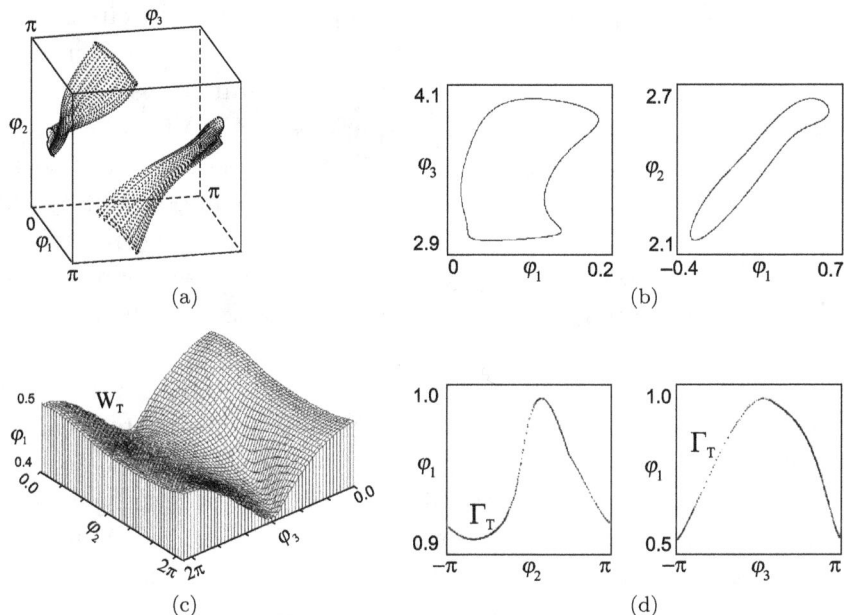

Fig. 4.6 Examples of $[0, 1, 1]$ rotation index attractor types: (a,c) three-dimensional projections of phase portraits and (b,d) two-dimensional projections of Poincaré maps.

$\varepsilon_1 = \varepsilon_2 = \varepsilon_3 = 1$ parameter values, determine the two-frequency quasisynchronous mode of the first generator and the two-frequency beat modes of the second and third generators. These attractors have similar $[0, 1, 1]$ rotation indices. However, they are not equivalent topologically. The first attractor represents a kind of a "tube". Its three-dimensional projection $(\varphi_3, \varphi_2, \varphi_1)$ is shown in Fig. 4.6(a). The two-dimensional projections, which show the attractor (φ_3, φ_1) cross sections with the $\varphi_2 = 0$ plane and (φ_2, φ_1) cross sections with the $\varphi_3 = 0$ plane, are shaped as a closed curve (Fig. 4.6(b)). The second attractor represents the surface depicted in Fig. 4.6(c), whose cross sections with the $\varphi_2 = 0$ and $\varphi_3 = 0$ planes are shaped as periodic curves with the period 2π (Fig. 4.6(d)).

Experiments with the models (4.2) and (4.3) have shown that invariant tube-shaped tori can be generated when a couple of limit

cycle complex-conjugate multipliers escape onto the unit circumference, i.e., in the Andronov–Hopf bifurcation. This bifurcation can
be soft. In this case, the loss of the limit cycle stability is accompanied by the generation of a stable invariant torus. Alternatively, it
can be hard when the change in the limit cycle stability is associated with the "constriction" of an unstable torus into a limit cycle.
In the latter case, the type of the newly appearing (disappearing)
attractor coincides with the type of the limit cycle, which undergoes
the Andronov–Hopf bifurcation, and is characterized by the rotation
index. For instance, the soft generation of the [0, 0, 0] rotation index
invariant torus (Fig. 4.4(b)) from the [0, 0, 0] rotation index limit
cycle can be observed as a consequence of the ε_2 parameter change
within the $\varepsilon_2 \in (0.07; 0.08)$ spectrum.

The second way of tube-shaped invariant tori generation in the V_6
phase space is connected with the saddle-node (saddle-node) bifurcation. In this connection, it must be noted that the rotation index of
the invariant torus generated from the saddle-node bifurcation will
not necessarily coincide with the limit cycle rotation index. Thus, in
the V_3 phase space of the model (4.3) at $\gamma_1 = 0.5, \gamma_2 = 0.5, \gamma_3 = 0.8,$
$\kappa_1 = 1.9$, and $\kappa_2 = 0.048$, there is a stable limit cycle, whose rotation
cycle is equal to [0, 0, 0]. Following the increase in the κ_2 parameter to 0.049, the oscillatory limit cycle disappears from a saddle-node
bifurcation, and a new [0, 0, 1] rotatory index invariant torus appears
in the V_3 phase space.

Numerical stimulation has revealed the only mechanism of
surface-shaped invariant tori generation — they are generated from
the limit cycle saddle-node bifurcation.

In the parameter space, G_{J_1, J_2, J_3} regions, whose region index coincides with the rotatory index of the attractor existing in the model's
phase space at the given region parameter values, correspond to partial quasisynchronous modes. The borders of the G_{J_1, J_2, J_3} regions
are formed by the curves, which correspond to the saddle-node
bifurcation of the limit cycle with $[J_1, J_2, J_3]$ rotation index and
to the formation of homoclinic paths encircling the phase cylinder
(torus) itself for φ_i coordinates, according to the applicable rotatory
index.

Mechanisms responsible for the excitation of quasisynchronous oscillations in the ensemble generators also present considerable interest. As a rule, excitation of self-oscillations in an individual generator is accompanied by the Andronov–Hopf bifurcation [38]. In the same way, self-oscillations can occur in an ensemble of coupled generators — that's exactly how they pass from the global synchronization mode to the global quasisynchronization mode. For a partial synchronization mode, excitation of self-oscillations in individual generators can follow different scenarios giving rise to a diversity of oscillation types.

Example 1. Let us consider the mode shown in Fig. 4.3(a), in which the first and the second generators are operating in the synchronous mode, while the third generator is operating in the beat mode. We will provide a non-zero feedback from the third generator to the second generator ($0 < \kappa_2 \ll 1$). As a result, oscillations from the third generator's local control circuit will be transmitted to the second generator's local control circuit and the latter will softly pass from the synchronization to the quasisynchronization mode. As for the first generator, it will still function in the synchronous mode due to $\kappa_1 \neq 0$.

The bifurcation diagram, which describes the self-oscillation excitation process in the second generator, is shown in Fig. 4.7(a). The vertical coordinate represents the ρ amplitude of the second generator phase shift dynamics $\rho = \max \varphi_2 - \min \varphi_2$ as compared to the reference generator. The horizontal coordinate represents the feedback quantity values. From the diagram, we can see that the oscillation amplitude depends on the feedback quantity — the higher κ_2 causes an increase in ρ. However, this dependency is not stable. Since the coupling is nonlinear, in certain circumstances, even a slight excursion can give rise to quantitative changes in the interacting generator's behavior. Figure 4.7(b) shows the attractor, which appears after the introduction of the feedback.

Example 2. Let us consider the motion shown in Fig. 4.3(b), when the first generator is operating in the synchronous mode, while the second and the third generators are functioning in the beat mode.

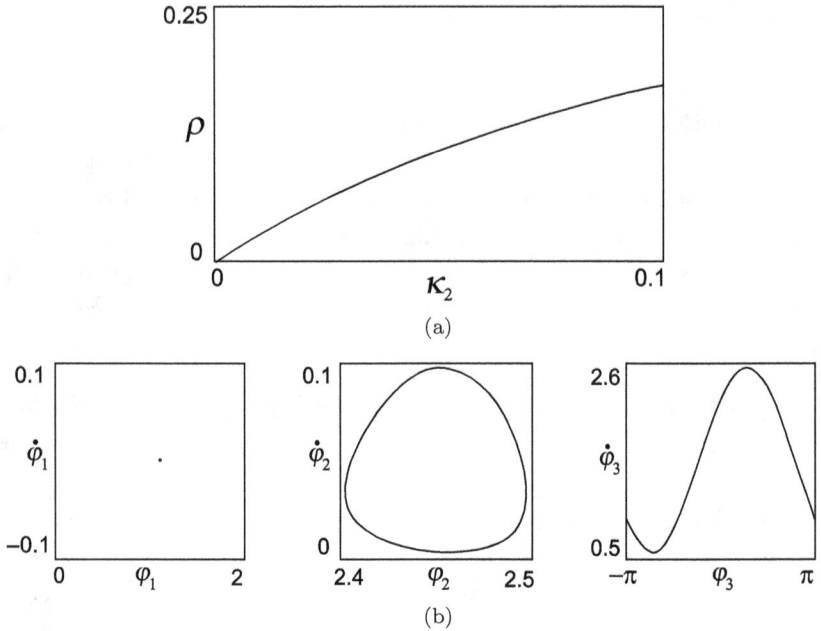

Fig. 4.7   Dependence of (a) the self-oscillations amplitude at the second CPLL generator output on the feedback quantity κ_2; (b) oscillatory–rotatory attractor of the model (4.3) at $\kappa_2 = 0.05$.

We will now provide a non-zero feedback from the second generator to the first generator $(0 < \kappa_1 \ll 1)$. As a result, the first generator will softly pass from synchronization to quasisynchronization mode. However, unlike the previous example, oscillations at the first generator's output have a more complicated nature. Further evolution of oscillations at the first generator's output caused by the increase in the κ_1 feedback parameter is shown in a single-parameter Poincaré map bifurcation diagram and phase portraits of attractors, as shown in Fig. 4.8. We can see that even a slight strengthening of the coupling force can change the nature of oscillations. To be more precise, under the influence of the saddle-node bifurcation, single-frequency oscillations turn into the two-frequency type. Calculation of Lyapunov exponents spectrum for the attractor presented in Fig. 4.8(c) confirms the presence of two-frequency oscillations. Lyapunov exponents calculated based on the $\tau = 10^5$ time interval have the following

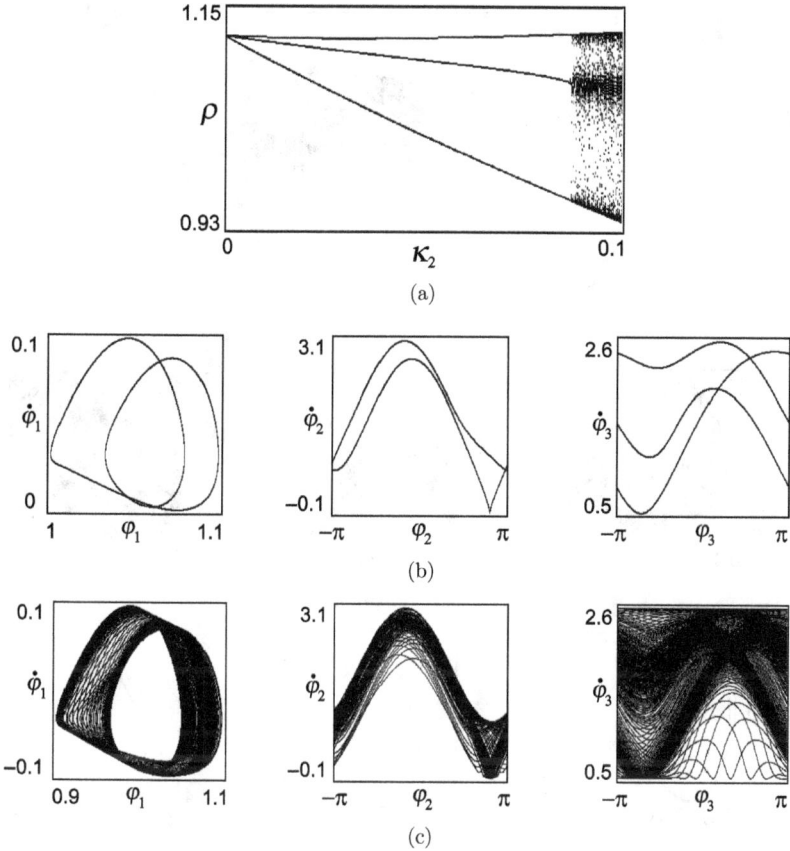

Fig. 4.8 (a) Poincaré map diagram and the model attractors (4.3) at (b) $\kappa_1 = 0.05$ and (c) $\kappa_1 = 0.1$, which characterize oscillations development at the first CPLL generator output following the change in the coupling force κ_1.

values: $\lambda_1 = 0.00001$, $\lambda_1 = -0.00002$, $\lambda_3 = -0.246672$. In other words, two of them are equal to zero, and the third has a negative value.

Example 3. Let us assume that the dynamic mode shown in Fig. 4.3(c) represents the initial system state. Now, by setting $\kappa_1 > 0$, we will immediately get two-frequency quasisynchronous oscillations at the first generator's output (Fig. 4.9(b)), which can be transformed into single-frequency oscillations (Fig. 4.9(c)) through further increase in κ_1.

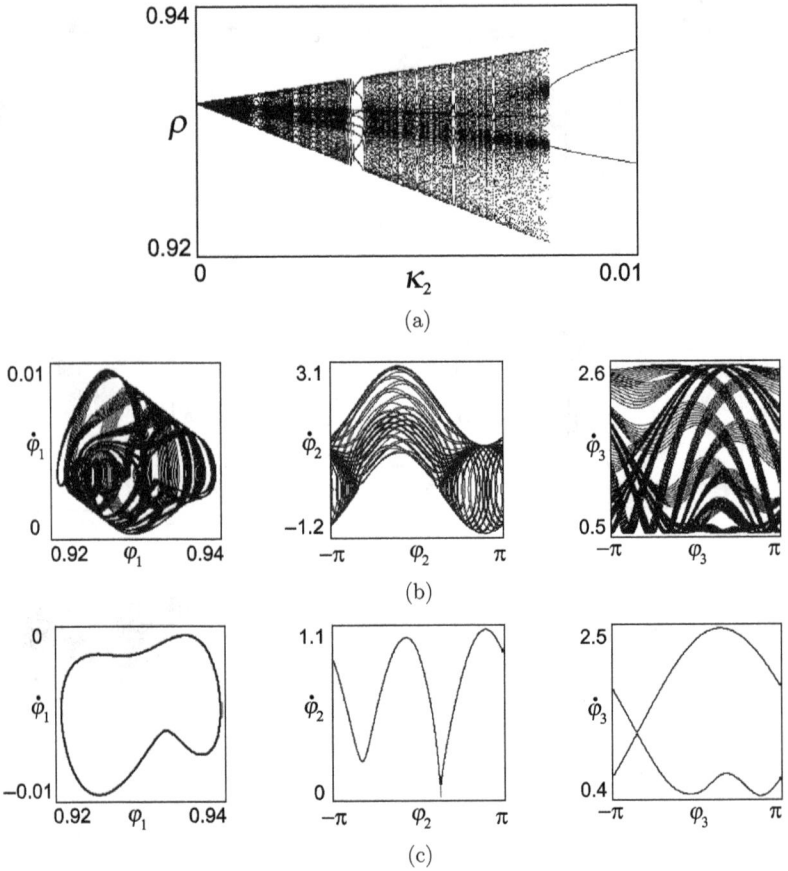

Fig. 4.9   (a) Poincaré map diagram and the model attractors (4.3) at (b) $\kappa_1 = 0.005$ and (c) $\kappa_1 = 0.01$.

In this way, the above examples demonstrate that if generators are joined in an ensemble, the transition from the synchronous mode to quasisynchronous oscillations can be achieved by the provision of additional couplings between partial subsystems. In such conditions, the excitation process takes on a soft character and is not accompanied by special trajectory bifurcations in the global phase spaces and the structure of the generated oscillations is determined by the oscillations in the control circuit of the generator.

4.4.   Bifurcation Transitions to Chaotic Oscillations

Chaotic attractors of the mathematical models (4.2) and (4.3) correspond to chaotic CPLL modes. Like regular attractors, chaotic attractors can encircle the V_6 (torus V_3) phase cylinder itself for all, some or none of the cylinder coordinates. The attractor type determines the type of oscillations at the output of controlled generators. We will identify the attractor type with the help of earlier introduced rotation indices. Now, the escape of phase trajectories onto a chaotic oscillatory attractor with the rotation index equal to $[0, 0, 0]$ will result in the generation of chaotically modulated oscillations (CMO) at the outputs of all ensemble generators. This is the CPLL *global CMO mode*. Finally, the generation of CMO at individual oscillators' output is associated with oscillatory–rotatory chaotic attractors, whose rotation indices have both zeros and unities. These attractors determine the partial generation mode in which a PLL ensemble generates CMO or a *partial CMO mode*. Last but not the least, the presence of a rotatory chaotic attractor with the rotatory index of $[1, 1, 1]$ is in line with the *global chaotic beat mode*.

Chaotic attractors appear in the CPLL models' phase spaces due to bifurcations undergone by regular attractors. In the following paragraphs, we will consider some scenarios of the occurrence and evolution of chaotic oscillations in CPLL models [64].

4.4.1.   *Transition to chaotic mode through a cascade of limit cycle period doublings*

Feigenbaum scenario is one of the key mechanisms for the excitation of chaotic oscillations in CPLL models. It is true for all motion types, in other words, it does not depend on the rotation index. Figure 4.10 illustrates the evolution of the oscillatory limit cycle when the κ_2 parameter increases from 0.63 to 2.35. Figure 4.10(a) shows a single-parameter bifurcation diagram of a Poincaré map, which characterizes the transition from a global regular quasisynchronous mode to the partial regular quasisynchronous mode through the global and

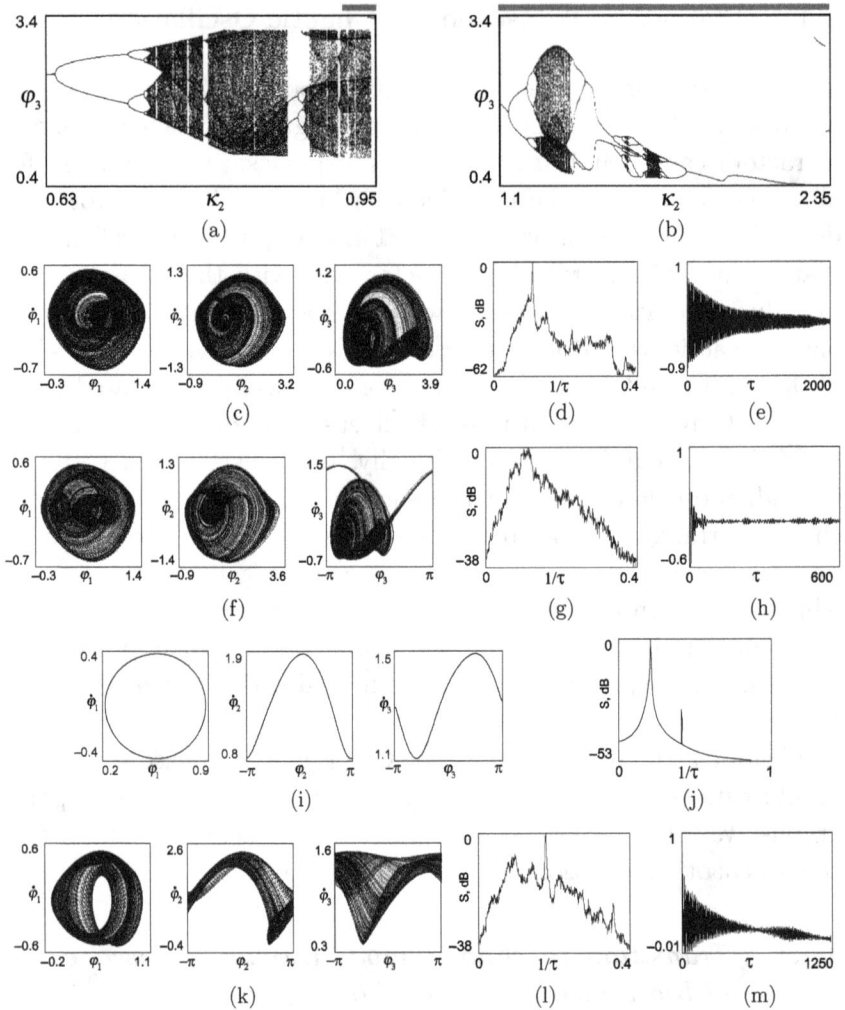

Fig. 4.10   (a, b) Fragments of Poincaré map single-parameter bifurcation diagram of the model (4.2) at $\gamma_1 = \gamma_2 = \gamma_3 = 0.5$, $\kappa_1 = 0.5$, $\varepsilon_1 = \varepsilon_2 = \varepsilon_3 = 1$; (c,d,e) projections, spectra and self-correlation functions of chaotic attractors: oscillatory at $\kappa_2 = 0.8$; (f,g,h) oscillatory–rotatory of the $[0, 0, 1]$ type at $\kappa_2 = 0.93$ and (i,j) of the $[0, 1, 1]$ type at $\kappa_2 = 1.3$; (k,l,m) $[0, 1, 1]$-type limit cycle projections and spectrum at $\kappa_2 = 0.95$.

partial CMO modes. The diagram in Fig. 4.10(b) illustrates CPLL's return to the global quasisynchronization mode by further strengthening of the κ_2 coupling force.

The limit cycle shown in Fig. 4.4(a) should be regarded as a starting state of the bifurcation diagram. With the increase in the κ_2 parameter, this limit cycle goes through a series of period-doubling bifurcations: at $\kappa_2^{(1)} = 0.638407$, $\kappa_2^{(2)} = 0.709835$, $\kappa_2^{(3)} = 0.724888$, $\kappa_2^{(4)} = 0.728068$, etc. These bifurcations generate a chaotic attractor with the $[0, 0, 0]$ rotation index in the model phase space (4.2). It determines the CPLL global CMO mode. An example of such chaotic attractor responsible for the generation of CMO by all ensemble generators is shown in Figs. 4.10(c–e). The value of the maximum Lyapunov exponent $\lambda_{\max} = 0.125$ confirms the chaotic character of this attractor.

Further strengthening of the κ_2 coupling results in the derangement of the third generator's quasisynchronous mode into a chaotic beat mode, whereas the first and the second generators continue functioning in the CMO generation mode. In this way, we can observe the establishment of a partial CMO mode in the ensemble. In the bifurcation diagram, the κ_2 parameter interval, in which the Poincaré map generated attractor rotates with respect to the φ_3 coordinate, is marked with a dark line above the diagram. The oscillatory–rotatory chaotic attractor with the $[0, 0, 1]$ rotation index, which is responsible for the generation of CMO at the first and second generators' outputs, is shown in Figs. 4.10(f–h) together with its spectrum and self-correlation function. Its maximum Lyapunov exponent has a positive value equal to $\lambda_{\max} = 0.138$.

Comparative analysis of the characteristics of oscillatory and oscillatory-rotatory attractors (Lyapunov exponents, spectra, self–correlation functions) shows that characteristics of CMO at the outputs of the first and second generators can be considerably modified by the variation of the third generator's dynamic mode through the variation of the second additional coupling force parameter. However, in the case under consideration, introduction of a strong coupling will cause the transition of the first generator to the single-frequency quasisynchronous oscillation mode, whereas the second and the third generators will change over to the single-frequency beat mode. The phase projections and the spectrum of the limit cycle, which corresponds to the regular partial quasisynchronization of the $[0, 1, 1]$-type CPLL, are shown in Figs. 4.10(i) and 4.10(j). This mode remains unchanged up to the $\kappa_2 = 1.1311$ values, when the $[0, 1, 1]$ limit cycle goes through a period-doubling bifurcation. Further increase in the κ_2 parameter is followed by more period-doubling bifurcations, which lead to the generation of a chaotic attractor with the rotation index equal to $[0, 1, 1]$. Figures 4.10(k–m) show the phase projections, the spectrum and the self-correlation function of the attractor characterized by the maximum Lyapunov exponent $\lambda_{\max} = 0.127$. This attractor accounts for the generation of CMOs exclusively at the output of the first ensemble generator.

Further increase in the κ_2 parameter results in the regularization of controlled generator oscillations within the intervals of (1.37; 1.57), (1.585; 1.64) and (1.7; 2.33167). At the κ_2 parameter values related to the said intervals, we can observe a $[0, 1, 1]$-type limit cycle in the V_6 phase space, while the partial regular quasisynchronization mode of the first generator is realized in the CPLL. Escape from the last interval through the $\kappa_2 = 2.33167$ limit is accompanied by the merging of the $[0, 1, 1]$ limit cycle into the O_5 saddle separatrix loop and transition of the system (4.2) to an oscillatory limit cycle. A global single-frequency quasisynchronous mode is established in the CPLL. Further increase in the κ_2 parameter transforms the previously established mode into a global synchronization mode I_6.

4.4.2. *Transition to chaotic mode through intermittence*

Let us simulate transition to chaotic PLCC oscillations through *Type I intermittency*, using the data on two cascade-coupled generators. We know that in a two-PLL ensemble model at $\gamma_1 = 0.7, \varepsilon_1 = 0$, $\gamma_2 = 0.3$ and variation of the κ_1 additional coupling force from 1.38 to 1.4, we will observe realization of the mechanism, which provides the transition to chaotic oscillations through Type I intermittency. In other words, we can say that the disappearance of a stable limit cycle brings the system to the chaotic state resulting from saddle-node bifurcation.

We will use this two-component cascade as the second and third CPLL generators with the $\kappa_2 = 1.38$ coupling. In doing so, we will choose the first generator parameters in the synchronization. In this way, we have constructed a CPLL system, which operates in the partial synchronization mode. Let us introduce a weak feedback from the second to the first generator, e.g., $\kappa_1 = 0.2$, to excite a quasisynchronous mode in the first generator. Now, the CPLL is operating in the partial quasisynchronous mode, which is matched by a stable limit cycle with the $[0, 1, 0]$ rotation index (Fig. 4.11) in the system phase space (4.2). By further increasing the κ_2 parameter, we can achieve the disappearance of the limit cycle. As a result, a chaotic attractor of the $[0, 1, 0]$ type will appear in the system phase space (4.2). The time development of the chaotic attractor phase variables near the bifurcation value of this parameter is characterized by the presence of long laminar regions interrupted by short chaotic outbursts (which is typical for intermittency related attractors). Further increase in the κ_2 parameter results in the reduction of the laminar sections and increase in the chaotic ones, and at $\kappa_2 = 1.416$, the behavior of phase variables becomes completely chaotic.

At $\gamma_1 = \gamma_2 = \gamma_3 = 0.5$, $\varepsilon_1 = \varepsilon_2 = \varepsilon_3 = 1$, $\kappa_2 = 1.8$ and $\kappa_1 \in (-0.9185, -0.9186)$, the system (4.2) demonstrates transition to chaotic oscillations through *Type II intermittency*. At $\kappa_1 = -0.9185$, a double limit cycle of the $[0, 1, 1]$ type with $\mu_{1,2} = 0.0000005 \pm i0.000004$, $\mu_3 = 0.000004$, $\mu_{4,5} = 0.1075 \pm i0.9904$ multipliers (Fig. 4.12(a)) exists in the V_6 phase space. Under the influence of the κ_1 parameter reduction to -0.9186, the $\mu_{4,5}$ multipliers become equal to $\mu_{4,5} = 0.1147 \pm i0.9947$, i.e., they leave the boundaries of the unit circumference. The limit cycle, which corresponds to the above multipliers, becomes unstable; phase trajectories escape from its neighborhood; however, some time later they return to the unstable limit cycle neighborhood just to escape from it again, etc. A chaotic attractor with a specific intermittency — long periods,

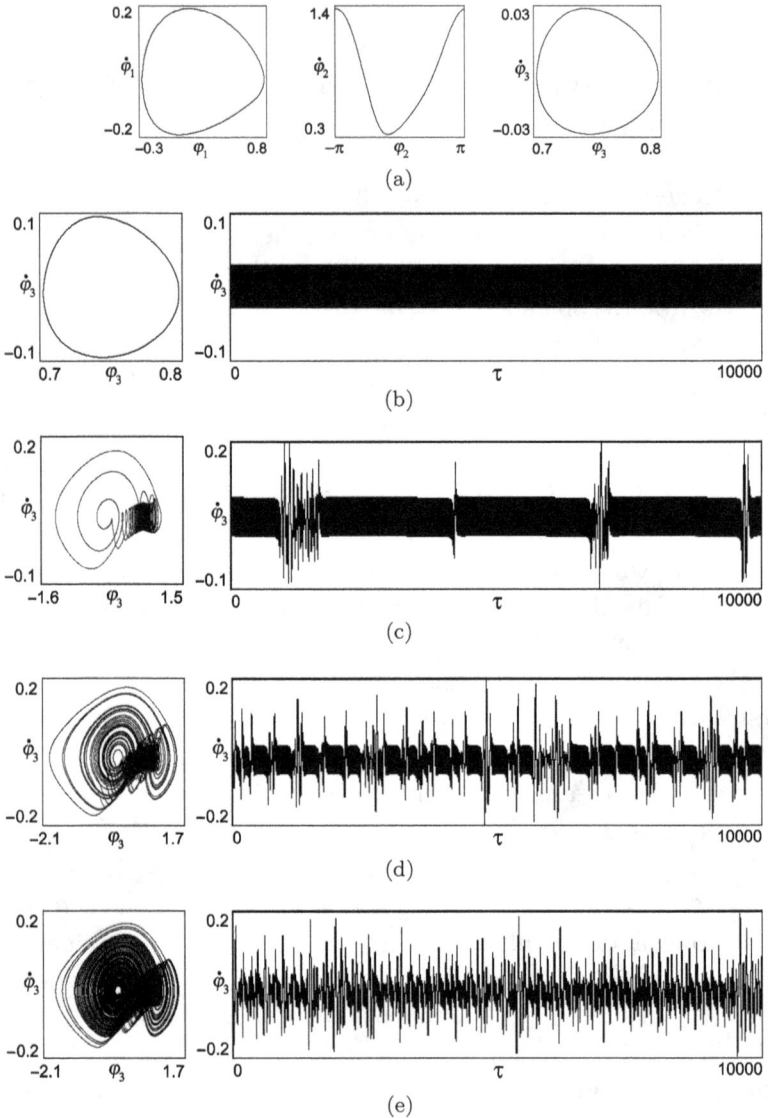

Fig. 4.11 Transition to chaotic oscillations through Type I intermittency: oscillatory–rotatory $[0, 1, 0]$-type limit cycle of the system (4.2) at (a) $\gamma_1 = 0.5, \varepsilon_1 = 1, \kappa_1 = 0.2, \gamma_2 = 0.7, \varepsilon_2 = 0, \gamma_3 = 0.3, \varepsilon_3 = 50, \kappa_2 = 1.39$, (φ_3, y_3) attractor projection and time realization of the y_3 variable at (b) $\kappa_2 = 1.39$, (c) $\kappa_2 = 1.39037$, (d) $\kappa_2 = 1.3915$, (e) $\kappa_2 = 1.416$.

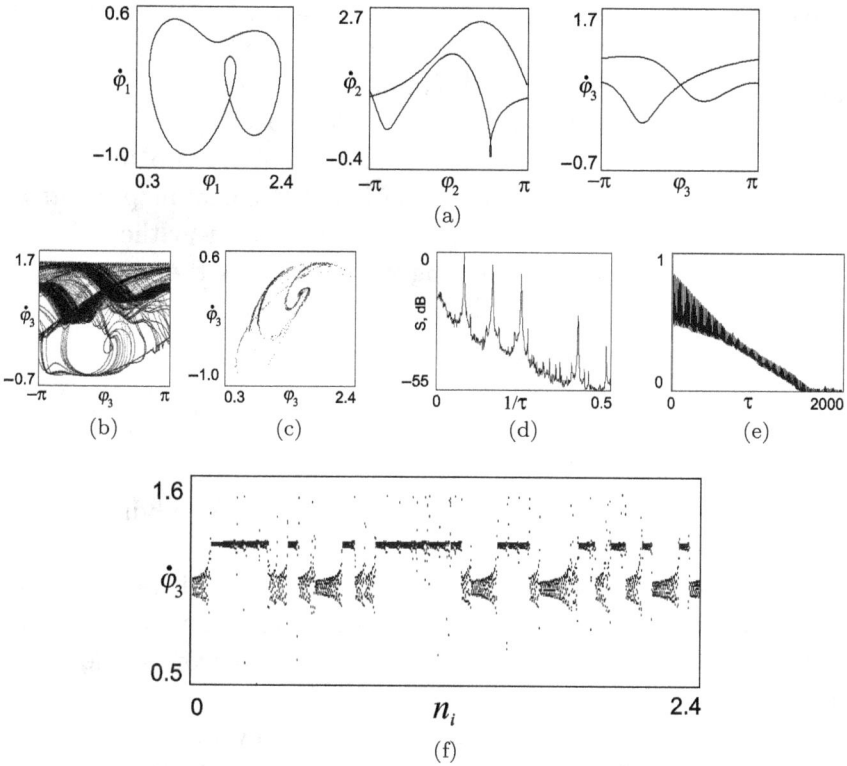

Fig. 4.12 Transition to chaotic oscillations through Type II intermittency: (a) attractor projections prior to bifurcation, (b) (φ_3, y_3)-projections of the phase portrait and (c) Poincaré map attractor after Andronov–Hopf bifurcation, (d) attractor's spectrum and (e) self-correlation function, (f) discrete-time realization of the y_3.

during which its phase trajectory runs within the unstable limit cycle neighborhood, are interrupted by short chaotic outbursts — appears in the phase space. Figures 4.12(b–e) show phase portrait projections and Poincaré maps, as well as the newly-generated chaotic attractor's spectrum and self-correlation function.

Intermittency properties of the above attractor can be illustrated with the help of Poincaré point map. We will now plot a diagram, in which the horizontal coordinate will display the number n_j of the point, at which chaotic attractor phase trajectory intersects with Poincaré secant; while the vertical coordinate will display the above point coordinate $y_3 = d\varphi_3/d\tau$. The result is shown in Fig. 4.12(b), in which we can clearly see the long period spent by the phase trajectory in the limit cycle neighborhood and short (up to 20 map points) random outbursts. Concentration of Poincaré map points at two levels observed in Fig. 4.12(f) can be explained by the fact that the unstable cycle is the doubling cycle, whereas the non-uniform distribution of the points between the two levels can be accounted for by the chaotic nature of the attractor.

4.4.3.   *Torus–chaos bifurcation*

Bifurcations of a stable two-dimensional torus can be subdivided into a proper torus-bifurcation, regarded as a stable invariant diversity, and bifurcation of torus-based periodic motions, which can lead to its destruction. Let us now consider the destruction of an invariant torus, which triggers chaotic attractor generation in the phase space.

Let us simulate the situation, when a stable invariant torus merges with the unstable one in the system phase space (4.2) leading to the generation of a chaotic attractor. For this purpose, we will have to construct a respective CPLL system. The above considered PLL cascade coupling, which demonstrates the transition to chaotic state through intermittency (see Sec. 3.5.3), will perform the functions of first and second generators, while the third generator will function in the beat mode. We will select the third generator's γ_3 and ε_3 parameters in such a way as to provide the generation of two-frequency oscillations when operating in the ensemble. Now, let us model the above-described system (4.2). Figure 4.13(a) shows an example of a stable invariant torus of the system (4.2) at $\gamma_1 = 0.7$, $\varepsilon_1 = 0$, $\gamma_2 = 0.3$, $\varepsilon_2 = 50$, $\gamma_3 = 0.5$, $\varepsilon_3 = 65$, $\kappa_1 = 1.38$, $\kappa_2 = 0$, which determines the single-frequency beat mode of the first generator, the single-frequency quasi-synchronous mode of the second generator and the two-frequency beat mode of the third generator. We would like

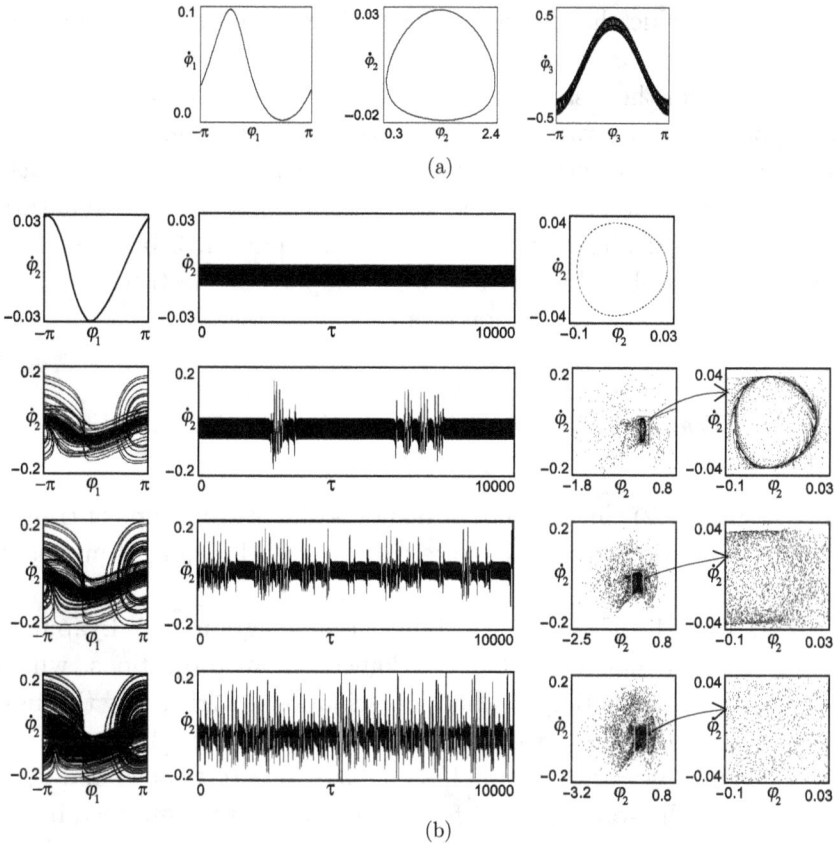

Fig. 4.13 (a) Stable invariant torus of the system (4.2) with rotatory index of $[1, 0, 1]$; (b) transition to chaotic oscillations through the "torus–chaos" type intermittency.

to draw your attention to the existence of an unstable limit cycle in the phase subspace $V_4 = \{\varphi_1 \pmod{2\pi}, y_1, \varphi_2 \pmod{2\pi}, y_2\}$. Characteristics of this limit cycle are very close to those of a stable limit cycle responsible for the generation of self-oscillations in the first two generators, which, in turn, cause two-frequency oscillations in the third generator. The above unstable limit cycle influences the third generator's periodic oscillations and accounts for the generation of an unstable invariant torus in the V_6 phase space. We will increase the

κ_1 coupling, which has previously been shown to cause the disappearance of limit cycles in a two-cascade system, and which will now be responsible for the disappearance of invariant tori caused by merging. Disappearance of invariant tori is accompanied by the generation of a chaotic attractor with the rotary index of $[1, 0, 1]$ in the V_6 phase space. Introduction of a weak κ_2 coupling will not change the nature of the observed bifurcation phenomenon. Figure 4.13(b) illustrates the above-described bifurcation, which implies the destruction of an invariant torus and generation of a chaotic attractor.

4.4.4. *Transition to chaotic mode through the doubling invariant tori*

The system (4.2) can demonstrate another interesting type of transition — transition to chaotic state through the doubling of invariant tori [60].

Figure 4.14 shows (a) the bifurcation diagram, (b) Lyapunov exponents dependence and (c–h) Poincaré map projections, which illustrate the transition to chaotic oscillations carried out through torus doubling. The invariant torus of the system (4.2) at $\gamma_1 = 1.7, \gamma_2 = 0.3, \gamma_3 = 0.5, \kappa_1 = 1.4, \kappa_2 = 0.35, \varepsilon_1 = 0, \varepsilon_2 = 50$ and $\varepsilon_3 = 65$ serves as the starting point for the plotting of the described bifurcation diagram. This attractor has the rotation index of $[1, 0, 0]$; its phase trajectory projectors are shown in Fig. 4.5(c), and its Poincaré maps are shown in Fig. 4.14(c). The increase in the ε_3 parameter generates bifurcations of period-2 torus (Fig. 4.14(d) at $\varepsilon_3 = 69$), of period-4 torus (Fig. 4.14(e) at $\varepsilon_3 = 70$), of period-8 torus (Fig. 4.14(f) at $\varepsilon_3 = 70.2$) and then generation of chaotic attractors (Fig. 4.14(g) at $\varepsilon_3 = 70.4$ and Fig. 4.14(h) at $\varepsilon_3 = 70.7$). Figure 4.14(b) represents dependencies of two maximum module Lyapunov exponents on the ε_3 parameter. We can see that two other exponents (accurate to the calculation parameters) are equal to zero and another exponent has a negative value. Invariant torus passage through the doubling bifurcation is accompanied by a factual zeroing of the maximum Lyapunov exponent, whereas the transition to chaotic attractor is followed by the said exponent's escape to the positive value region.

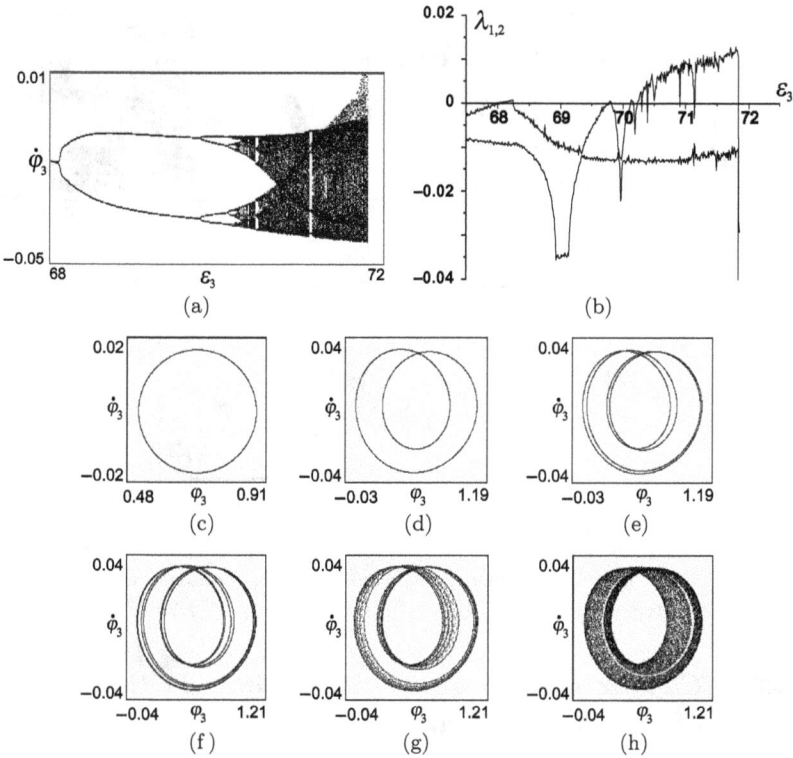

Fig. 4.14 Transition to chaotic oscillations through the two-dimensional torus doubling bifurcation: (a) single-parameter bifurcation diagram; (b) dependence of two maximum module Lyapunov exponents on the ε_3 parameter; (c–h) two-dimensional $(\varphi_3, \dot{\varphi}_3)$-projections of Poincaré map.

At $\varepsilon_3 \approx 71.84$, the chaotic attractor is destroyed and phase trajectories from its neighborhood move on to the $[1, 1, 1]$-type invariant torus. Phase portrait projections of this rotation torus together with the projections of Poincaré section by the $\varphi_1 = 0$ plane are shown in Fig. 4.15. Interestingly, although Poincaré section takes the form of closed curves when projected to the (φ_2, y_2) plane, and of periodic curves when projected to the (φ_3, y_3) plane, the rotation of the phase trajectories is observed for both φ_2 and φ_3 coordinates and the global two-frequency beat mode is established throughout the CPLL system.

(a)

(b)

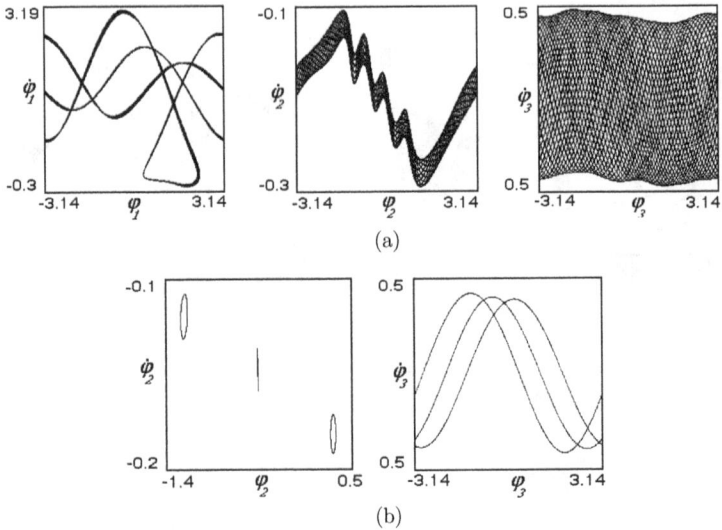

Fig. 4.15   (a) Phase trajectories projections and (b) Poincaré maps of a two-dimensional torus with the $[1,1,1]$ rotation index at $\varepsilon_3 = 71.85$.

4.4.5.   *Chaotic mode provoked by the introduction of additional couplings*

In Sec. 4.3, we examined how individual generators, which form part of the ensemble, can be switched over from synchronous mode to various regular quasisynchronous modes by introducing new additional couplings by using mismatching signals. In the same way, they can be switched over from synchronous mode to CMO generating mode.

This requires giving a signal from the PLL_2 or PLL_3 control circuits, which function in the chaotic mode, to the local PLL_1 control circuit, which functions in the synchronous mode. Figure 4.16(a) shows projections of the attractor responsible for the generation of CMO at the first and third generator's outputs and chaotic beats at the second generator output. Figures 4.16(b) and 4.16(c) show the power spectrum and the self-correlation function of the generated chaotic attractor. Chaotic oscillations at the first generator output have been generated through the introduction of an additional coupling characterized by the κ_1 parameter.

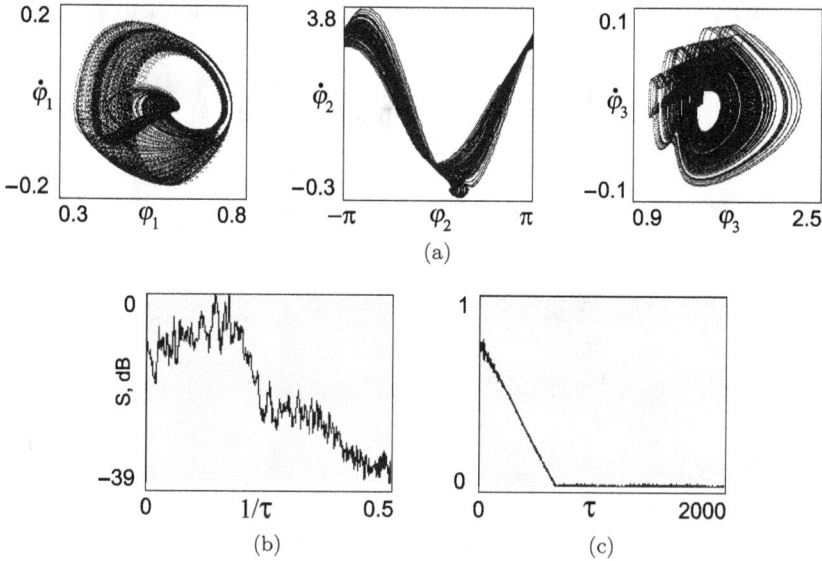

Fig. 4.16 (a) Projections, (b) spectrum and (d) self-correlation function of the $[0, 1, 0]$-type model chaotic attractor (4.2) at $\gamma_1 = 0.5$, $\varepsilon_1 = 1$, $\gamma_2 = 1.7$, $\varepsilon_2 = 0$, $\gamma_3 = 0.68$, $\varepsilon_3 = 50$, $\kappa_1 = 0.2$, $\kappa_2 = 1.9$.

4.5. Asynchronous Modes

It is most interesting that when quasisynchronous oscillations are generated through the introduction of an additional coupling, generators, which serve as the source of modulating oscillations, can function both in the quasisynchronous and beat modes. Because of this, previously, parasitic beat modes were most efficiently used for the generation of quasisynchronous oscillations of varying complexity. Let us study an example to illustrate the possibility of the formation of quasisynchronous oscillations of varying nature with the help of the beat mode.

For this purpose, we will construct a system consisting of four cascade-coupled PLL systems by connecting the CPLL (formed by three PLLs) ensemble under consideration of the PLL$_0$ system provided with a first-order filter. Then we will introduce additional couplings through control circuits between PLL$_0$ and local control circuits of all CPLL generators. In this case, dynamic processes in the newly constructed cascade will be described by the following mathematical model:

$$\frac{d\varphi_0}{d\tau} = y_0, \quad \varepsilon_0 \frac{dy_0}{d\tau} = \gamma_0 - y_0 - \sin\varphi_0 - \kappa_{01}\sin(\varphi_1 - \varphi_0)$$

$$-\kappa_{02}\sin(\varphi_2 - \varphi_1) - \kappa_{03}\sin(\varphi_3 - \varphi_2),$$

$$\frac{d\varphi_1}{d\tau} = y_1, \quad \varepsilon_1 \frac{dy_1}{d\tau} = \gamma_1 - y_1 - \sin(\varphi_1 - \varphi_0) - \kappa_1\sin(\varphi_2 - \varphi_1),$$

$$\frac{d\varphi_2}{d\tau} = y_2, \quad \varepsilon_2 \frac{dy_2}{d\tau} = \gamma_2 - y_2 - \sin(\varphi_2 - \varphi_1) - \kappa_2\sin(\varphi_3 - \varphi_2),$$

$$\frac{d\varphi_3}{d\tau} = y_3, \quad \varepsilon_3 \frac{dy_3}{d\tau} = \gamma_3 - y_3 - \sin(\varphi_3 - \varphi_2),$$

$$(4.5)$$

where φ_0 and y_0 are phase variables, which characterize the state of the first generator for the reference signal; γ_0, ε_0 are the first PLL parameters, while κ_{01}, κ_{02} and κ_{03} are additional coupling parameters, which characterize the connections formed by the first generator with the second, third and fourth generators respectively.

We will now select CPLL parameter values, which reflect the realization of the global beat mode, e.g., the two-frequency oscillatory mode, which is matched by the rotatory invariant torus (Fig. 4.15) in the model phase space (4.2). Let us set the PLL$_0$ parameters equal to $\gamma_0 = 0.1, \varepsilon_0 = 5$. In other words, these parameters shall ensure that in the absence of additional couplings, the PLL$_0$ system should operate in the synchronous mode. Then we will try to receive diverse quasisynchronous oscillations at the PLL$_0$ output by modulating only the $\kappa_{01}, \kappa_{02}, \kappa_{03}$ coupling parameters. Figure 4.17 shows projections of Poincaré maps and $\varphi_0(\tau)$ variable realizations at different coupling parameter values as examples of the achieved quasisynchronous oscillations. Figures 4.17(a) and 4.17(c) illustrate two-frequency quasisynchronous oscillations modes; Fig. 4.17(b) illustrates the single-frequency quasisynchronous oscillations mode; and Fig. 4.17(d) illustrates the CMO generation mode. In this way, from the above example, we can see that oscillations in the control circuits of PLL systems functioning in beat modes can be efficiently used for the generation of quasisynchronous oscillations with varying characteristics.

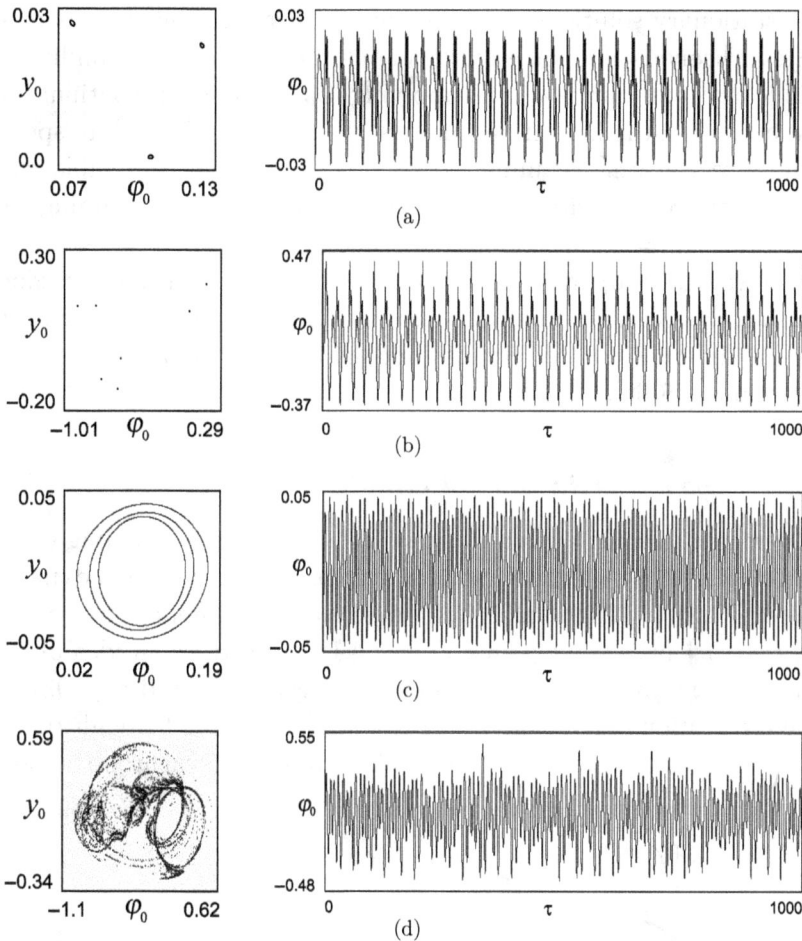

Fig. 4.17 Projections of Poincaré map and the $\varphi_0(\tau)$ variable realization of the model (4.5) at (a) $\gamma_1 = 1.7, \varepsilon_1 = 0, \gamma_2 = 0.3, \varepsilon_2 = 50, \gamma_3 = 0.5, \varepsilon_3 = 71.85, \kappa_1 = 1.4, \kappa_2 = 0.35, \gamma_0 = 0.1, \varepsilon_0 = 5, \kappa_{01} = 0.1, \kappa_{02} = \kappa_{03} = 0$; (b) $\kappa_{01} = 0, \kappa_{02} = 0.8, \kappa_{03} = 0$; (c) $\kappa_{01} = \kappa_{02} = 0, \kappa_{03} = 0.1$; (d) $\kappa_{01} = 0.4, \kappa_{02} = 0.2, \kappa_{03} = 0.5$.

4.6. Analysis of the Parameter Space Structure

High dimensionality of CPLL models, the presence of three cyclic coordinates and a large number of various parameters complicate the full breakdown of the parameter space into individual regions characterized by specific dynamic behavior. That is why, when examining the parameter space structure, we will, firstly, focus exclusively on the stationary solutions that present real practical interest, namely on synchronous and quasisynchronous motions, and, secondly, we will carry out our research applying the parameter continuation method using the known parameters of simpler models. More specifically, we will first examine the dynamics of a PLL ensemble with low-inertia control circuits at similar initial frequency detuning values $\gamma = \gamma_1 = \gamma_2 = \gamma_3$ and feedback couplings $\kappa_1 = \kappa_2 = \kappa$. Then we will study synchronous and quasisynchronous mode regions for $\kappa_1 \neq \kappa_2$, and, finally, consider dynamic behavior peculiarities at different control circuits and additional coupling parameter variations [65, 66].

4.6.1. *Low-inertia control circuits*

At low-inertia control circuits, equal initial frequency detuning values and similar feedback couplings, the system (4.2) turns into a dynamic system (4.3) with one-and-a-half degrees of freedom identified on the $\Lambda_3 = \{\gamma, \kappa\}$ parameter plane. Figure 4.18 can give us an idea of this plane. At negative feedback coupling values $\kappa < 0$ and $\gamma \geq 0$, only one bifurcation curve $\gamma^+(\kappa) = (1 - \kappa + \kappa^2)^{-1}$ passes through the Λ_3 plane. This curve separates the I_1 globally stable synchronous mode

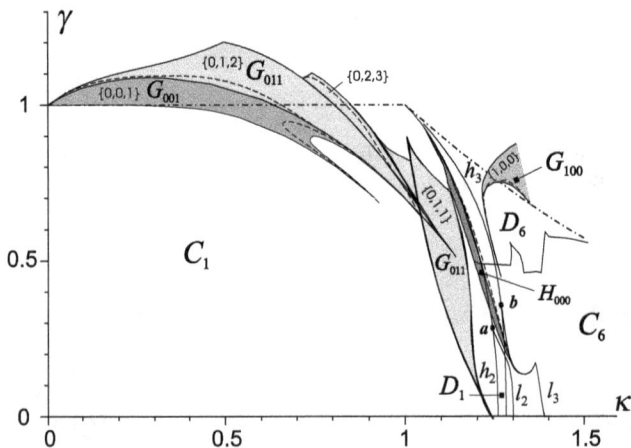

Fig. 4.18   Parametric portrait of the system (4.3) at $\kappa_1 = \kappa_2 = \kappa$, $\gamma_1 = \gamma_2 = \gamma$.

existence region from the asynchronous mode existence region. The structure of this plane sector is extremely simple. That is why it is not shown in Fig. 4.18. Since the system (4.3) is invariant for the replacement of $(\gamma, \kappa, \varphi_1, \varphi_2, \varphi_3)$ for $(-\gamma, -\kappa, -\varphi_1, -\varphi_2, -\varphi_3)$, it will be enough to examine the $\gamma \geq 0$ values.

In Fig. 4.18, the γ^+ dotted line restricts the C_0 region of the model (4.3) stability state existence. Within the C_0, we can see subregions $C_1 = \{0 < \gamma < \min(\gamma^+, h_2)\}$ and $C_6 = \{\min(0, h_3) < \gamma < \gamma^+\}$, which are existence regions of the I_1 and I_6 synchronous modes, respectively. No synchronous modes are observed between the h_2 and h_3 curves.

The h_2 curve represents the loss of stability undergone by the O_1 equilibrium state, which corresponds to the I_1 synchronous mode. The a point located on this curve divides the examined border-line into the safe and dangerous sections. When leaving the C_1 region through the h_2 curve section located below the a point, the O_1 equilibrium state undergoes a soft loss of stability accompanied by the generation of a stable low-amplitude oscillatory limit cycle L_{01}. Increase in the κ, γ parameters causes growth of the limit cycle amplitude, and when the system parameters reach the l_2 curve, the L_{01} cycle merges with the saddle cycle and disappears. In this way, we can observe realization of a global quasisynchronous mode in the $D_1 = \{h_2 < \kappa < l_2\}$ region of the CPLL phase space. This mode is the only one found in the $\overline{D}_1 = \{h_2 < \kappa < \min(l_2, h_3)\}$ region, whereas in the $\widetilde{D}_1 = D_1 \setminus \overline{D}_1$ region, it exists together with the I_6 synchronous mode. When leaving the C_1 region through the curve section located above the a point, the equilibrium state undergoes a hard loss of stability, which is accompanied by the CPLL switchover to asynchronous modes.

Like the h_2 curve, the h_3 curve, which represents the loss of stability undergone by the O_3 equilibrium state responsible for the set-in of the I_6 synchronous mode, is divided into two sections by the b point. In the h_3 section located below the b point, the first Lyapunov exponent has a positive value. When this section is crossed with the reduction of the κ (or γ) parameters, the change in the I_6 mode stability takes a hard turn. From this bifurcation, the CPLL switches over to the global regular quasisynchronous mode determined by a stable oscillatory cycle L_{03}. Further reduction of the κ (or the γ) parameter results in the L_{03} period-doubling bifurcations, which lead to the generation of the S_{03} chaotic oscillatory attractor in the V_3 phase space. This attractor determines the global CMO mode. In Fig. 4.18, the H_{000} region corresponds to the mode, in which all ensemble generators can generate CMO. The $\overline{H}_{000} \subset H_{000}$ subregion located between the h_2 and h_3 curves is the region of the S_{03} attractor global stability, where the global CMO mode sets in by default. On leaving the H_{000} region through the left border, the S_{02} attractor breaks down. The mode, to which the CPLL switches over following escape from the H_{000} region through the left border, depends on the γ parameter value. It can be either a synchronous I_6 mode (in case the border is crossed to the left of the h_3 curve), or a partial quasisynchronization, synchronization or asynchronous mode. The l_3 bifurcation curve, on which one of the l_{03} limit cycle multipliers takes the $+1$ value adjoins the h_3 curve at b.

At the parameter values from the G_{001} region, we can observe attractors with the $[0, 0, 1]$ rotation index in the V_3 phase space, and a mode, where the first and the second generators generate quasisynchronous oscillations, settles in the CPLL system. The upper boundary of this region is determined by the saddle-node bifurcation of the limit cycles, while the lower boundary is determined either by the loop of the O_3 separatrix saddle (saddle-node) with a negative saddle value, or by a saddle-node bifurcation. The right boundary of the G_{001} region has a complex structure: it includes bifurcation curves of multiturn separatrix loops, saddle-node bifurcations of the multifold cycles and the line corresponding to the crisis of a chaotic attractor with the $[0, 0, 1]$ rotation cycle.

At the parameter values from the G_{011} region, only the first generator can function in the quasisynchronous mode. In the system phase space (4.3), this mode is matched by the attractors limited only for one cyclic coordinate φ_1. We would like to remark that this mode is matched not just by one limit cycle or chaotic attractor, due to which it is generated, but by a whole set of cycles. Attractors belonging to this set can be distinguished by the number of revolutions around

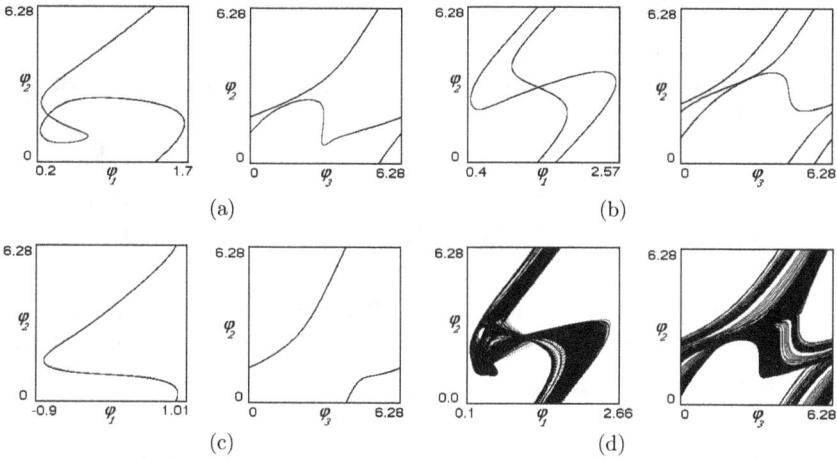

Fig. 4.19 Examples of various types of the model attractors (4.3) responsible for the establishment of quasisynchronous oscillations at the output of the first generator: regular at (a) $\gamma = 0.9$, $\kappa = 0.8$, (b) $\gamma = 1.09$, $\kappa = 0.75$, (c) $\gamma = 0.11$, $\kappa = 1.2$ and chaotic (d) $\gamma = 0.99, \kappa = 0.68$.

the V_3 torus, which are performed by a cycle limit point with respect to the φ_2 and φ_3 coordinates before closure. In Fig. 4.19, we can see various system oscillatory–rotatory attractors (4.3) responsible for the first generator's partial synchronization mode. In Fig. 4.18, the partial synchronization region G_{011} is divided into subregions, which correspond to different types of stationary motions. These motions are characterized by the number of V_3 phase torus encircling (revolutions made around the phase torus) with respect to the φ_1, φ_2 and φ_3 coordinates by the "basic" limit cycles ("basic" limit cycles are cycles before period doubling).

In the phase space of the system (4.3), we can observe phase trajectories for the parameter values from the G_{100} region, which determine quasisynchronous oscillations at the outputs of the second and third generators. When leaving this region through its upper boundary, the $\{1, 0, 0\}$ limit cycle disappears if one of the multipliers is equal to $+1$. When leaving this region through its right boundary, the stable limit cycle disappears as the couple of the complex-conjugate multipliers reaches the unit circumference. Escape through the bottom boundary is accompanied by a cycle period-doubling cascade, generation of a chaotic attractor and its eventual destruction. Figures in curly brackets characterize the rotation number of the "basic" limit cycle. Dotted lines in quasisynchronization regions correspond to the first bifurcations of the basic cycle's period doubling. Escape from the G_{100} region always leads to the establishment of the CPLL asynchronous modes irrespective of the border, through which it is affected.

The parametric portrait $\{\kappa, \gamma\}$ shown in Fig. 4.18 explains the non-trivial behavior of ensemble generators at the variation of the coupling and initial frequency detuning. From Fig. 4.18, we can see that parameter regions, in which generators can generate quasisynchronous oscillations, have extremely irregular shapes and are not necessarily singly-connected. It happens because quasisynchronous regions are formed by a merger of several existence regions of various attractor types. Thus, the quasisynchronous oscillation existence region at the first generator's output unites D_1, G_{001} and G_{011} regions.

We would like to remark that the G_{011} region is also composite. It accommodates existence regions of limit cycles characterized by various rotation numbers and of more complicated attractors formed on the basis of the said cycles. Since chaotic attractors with the rotation index of $[0, 1, 1]$ can be formed based on limit cycles with unequal rotation numbers, the CMO generation H_{011} region at the first generator's output will be multiply-connected even at $G_{0,1,1}$ region parameter values. Interestingly, existence regions of different types of attractors can overlap, causing instability of quasisynchronous oscillations at the CPLL system parameter variation.

In conclusion, we will compare the received parametrical portrait shown in Fig. 4.18 with a similar portrait shown in Fig. 3.12 for the model (3.5) of two cascade-coupled generators. This comparison will enable us to detect changes in the ensemble dynamics caused by the increase in the number of coupled PLLs. The system of three identical cascade-coupled generators with low-inertia control circuits, as well as a system of two similar generators, can operate in two synchronous modes. One of them (the I_1 mode) can be observed at "weak" connections, and the other (the I_2 modes and the I_6 mode for three coupled generators) at "strong" connections. As the number of the elements increases, existence regions of synchronous modes undergo reduction and synchronous mode stability change curves move to the κ region of lesser values, i.e., synchronous modes change their stability at weaker connection values. If the change in synchronous mode stability took place in two cascade-coupled generators at the same time, the change in the stability of the I_1 and I_6 synchronous modes in a three-system ensemble will be observed at varying κ parameter values. This phenomenon leads to the formation of a region within the parameter region of three coupled systems, where the ensemble does not generate any synchronous motions at all. In two cascade-coupled generators, the change in the I_1 mode stability always took a hard turn, whereas the change in the I_6 mode stability was always of a softer nature. In a three-generator ensemble, the change in the stability of both the I_1 and I_6 modes can be soft, as well as hard. There is a possibility to generate chaotic oscillations from a crucial new phenomenon observed within an ensemble of three phase-controlled generators with low-inertia control circuits. Another interesting phenomenon that attracted the authors' attention was one more type of motion detected in the model of three identical PLL chain. These motions belong to the $[1, 0, 0]$ type, where the first generator operates in an asynchronous mode for the reference signal, while the remaining two generators function in the quasisynchronous mode. The latter is interesting by the fact that the second generator, for which the first generator's output oscillations serve as an input signal, operates in the quasisynchronous mode for the reference signal. This situation is impossible in an ensemble consisting of two identical PLLs, as in the phase space of the model (3.5) at $\gamma_1 = \gamma_2 = \gamma$ and $\delta = 0$ $[1, 0]$-type motions are not observed at any γ and κ values. A dynamic mode of the $[1, 0]$ type in an ensemble consisting of two cascade-coupled PLLs is attainable only for phase systems with non-identical parameters, i.e., at $\gamma_1 \neq \gamma_2$.

4.6.2.   *Non-identical additional couplings*

In this section, we will try to answer the following question: "What happens to ensemble dynamic modes and their existence regions in the parameter space following the violation of additional coupling identity/similarity?" For this purpose, we are going to analyze two-parameter portraits of the model (4.3) on the (κ_1, κ_2) plane plotted for varying initial frequency detuning values: $\gamma = 0.3$ (Fig. 4.20), $\gamma = 0.5$ (Fig. 4.21) and $\gamma = 0.7$ (Fig. 4.22).

In Figs. 4.20–4.22, dotted lines form borders of the equilibrium states' existence region C_0. Lines h_1, h_2, h_3, h_4 correspond to Andronov–Hopf bifurcations and divide the C_0 region into the C_1,

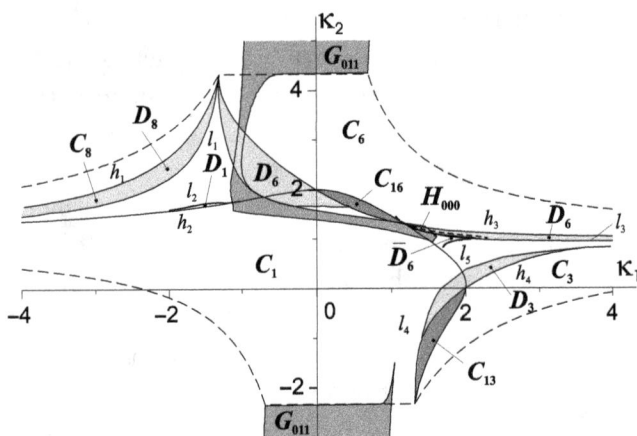

Fig. 4.20   The model (4.3) dynamic mode diagram at $\gamma_1 = \gamma_2 = \gamma_3 = 0.3$.

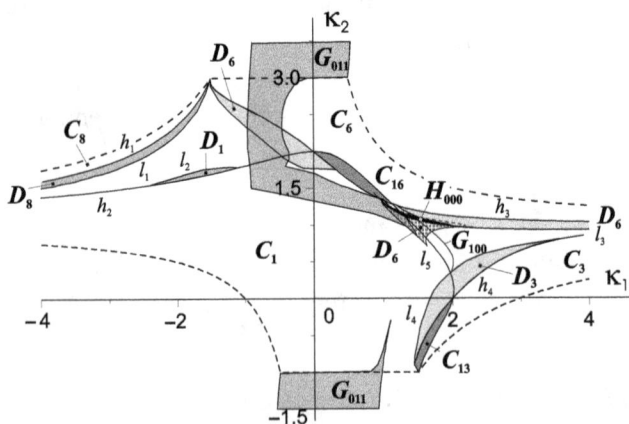

Fig. 4.21   The model (4.3) dynamic mode diagram at $\gamma_1 = \gamma_2 = \gamma_3 = 0.5$.

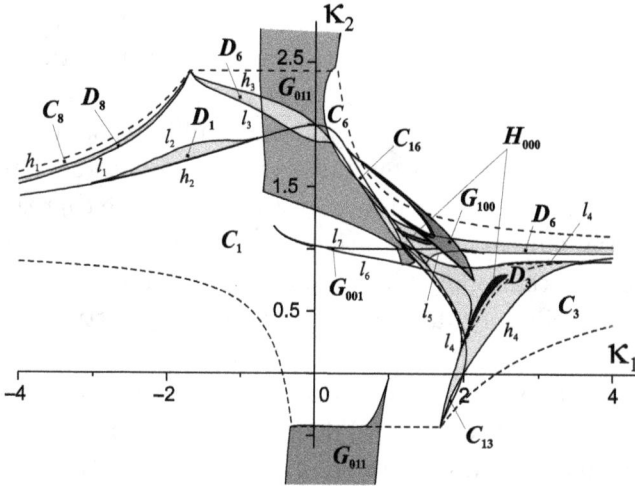

Fig. 4.22   The model (4.3) dynamic mode diagram at $\gamma_1 = \gamma_2 = \gamma_3 = 0.7$.

C_3, C_6 and C_8 subregions, where the O_1, O_3, O_6 and O_8 equilibrium states are stable and consequently serve as regions of the I_1, I_3, I_6 and I_8 synchronous modes realization. Two synchronous modes, which differ by second and the third generator synchronization phase errors, can be found in the $C_{16} = C_1 \cap C_6$ and $C_{13} = C_1 \cap C_3$ regions. The letter D in the (κ_1, κ_2)-diagrams denotes existence regions of global single-frequency quasisynchronous CPLL modes.

The borders of the D_1, D_6 and D_8 regions are outlined by the h_1, h_2, h_3 lines and the l_1, l_2, l_3 separatrix loop bifurcation curves adjacent to Andronov–Hopf bifurcation curves at the points of the first Lyapunov exponent sign change. The borders of the $\bar{D}_6$ and D_3 regions alongside the h_3 and h_4 curves include the l_5 and l_4 lines composed of saddle-node bifurcation curves and of the curves responsible for the separatrix loops formation. It should be noted that oscillatory attractors and, consequently, CPLL global quasisynchronism modes can be realized only at the C_0 region parameter values.

The $G_{J_1 J_2 J_3}$ regions correspond to partial quasisynchronization modes, where the region index coincides with the rotation index of the respective attractor, which exists in the model phase space at the said region's parameter values. The borders of the $G_{J_1 J_2 J_3}$ regions are outlined by the curves, which correspond to the saddle-node bifurcation of the limit cycle with the rotation index equal to $[J_1, J_2, J_3]$ and to the bifurcation of homoclinic trajectories encircling the phase torus itself with respect to the φ_i coordinates in accordance with the respective rotation index. Unlike existence regions of attractors with other rotation index values, the existence regions of $[0, 0, 0]$-type attractors are always located within the C_0 region. That is why partial quasisynchronization mode existence regions are much more extensive than global synchronization regions.

Regular oscillatory and oscillatory–rotatory attractors of the model (4.3) at $\kappa_1, \kappa_2, \gamma$ parameter variations can turn chaotic. In most cases, it happens in accordance with the Feigenbaum scenario. In Figs. 4.20–4.22, the letter H denotes existence regions of chaotic oscillatory and oscillatory–rotatory attractors. In this case study, region indices coincide with respective attractors' rotation indices.

The analysis of Figs. 4.18 and 4.20–4.22, as well as the results of other computing experiments carried out with the model (4.3) show that by matching respective connection parameters, the generators can be switched over to CMO mode even for low-inertia control circuits. However, in this case, chaotization of quasisynchronous motions has been detected only at $\gamma > 0.3$. It has been established that the diversity of quasisynchronous modes grows with the increase in the γ parameter. From Figs. 4.18 and 4.20–4.22, we can see that at low γ ($\gamma \leq 0.3$) values, we can observe regular motions with rotation indices equal to $[0, 0, 0]$ and $[0, 1, 1]$. Increase in the γ values leads to the formation of parameter regions with the motions of the $[0, 0, 1]$ and $[1, 0, 0]$ types, chaotization of oscillatory and oscillatory–rotatory motions and merger of different types of attractors, which, in turn, results in the generation of attractors of a more complicated structure. Within the framework of the above processes, non-identical connections allow much more room for controlling synchronous and quasisynchronous mode properties.

4.6.3. *Ensemble with inertia control circuits*

This section contains the results of the studies carried out about motions typical of a CPLL model with inertia control circuits.

The main focus is placed on quasisynchronous modes. The authors describe CMO existence regions in various parameter space cross sections and analyze the inner structure of the specified regions and modulating oscillation properties.

The analysis of the CMO generation regions within the model (4.3) parameter space proves that in a CPLL system with low-inertia control circuits these regions are limited. However, the studies of the model (4.2) chaotic oscillations show that these regions can be considerably expanded by the introduction of inertia into PLL local control circuits.

Figure 4.23 illustrates the development of the ensemble quasisynchronous mode's existence regions caused by the increase in the PLL local control circuit inertia. We can see the same symbols and signs as those used for the denotation of regions and curves in Fig. 4.21. From the comparative analysis of bifurcation diagrams in Figs. 4.23 and 4.21, it follows that a small increase in the control circuit inertia value leads to significant changes in the ensemble dynamics. In the first place, it shows up in the reduction of synchronous mode existence regions and, as a consequence, in the disappearance of bistable synchronous behavior regions. Quasisynchronous modes with new quasisynchronism indices (not discovered at $\varepsilon = 0$) appear in the

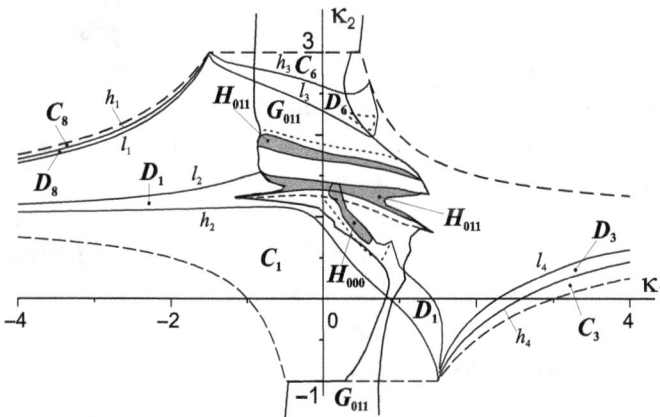

Fig. 4.23 The model (4.2) dynamic mode diagram at $\gamma_1 = \gamma_2 = \gamma_3 = 0.5$, $\varepsilon_1 = \varepsilon_2 = \varepsilon_3 = 1$.

phase space of the model (4.2). As the ε parameter increases, quasisynchronous mode existence regions expand and begin to overlap each other causing multistable behavior. As for quasisynchronous modes, they become more complex due to period-doubling bifurcations. Meanwhile, chaotic attractors with different rotation indices can merge giving a start to even more complex motions. In Fig. 4.23, some bifurcation curves, which correspond to the first cycle period-doubling bifurcations, are shown with dotted lines. These curves can be regarded as assessment of the CMO existence regions. It should be noted that increase in the control circuit inertia causes basic changes in the ensemble dynamics at small ε values. Further increase in the ε parameter results in the stabilization of the parameter space structure.

This conclusion is true both for a homogeneous chain, when the ensemble is composed of identical components, and for a model comprising different components. This statement is illustrated by Fig. 4.24(a), which shows a CMO existence region at the output of the first CPLL generator. In the phase space of the model (4.2), these oscillations are matched by chaotic attractors, whose quasisynchronism indices are equal to $[0, 0, 1]$ and $[0, 1, 1]$. The single-parameter bifurcation diagram of Poincaré map $\{\varepsilon_1, \varphi_1\}$ (Fig. 4.24(b)) and Lyapunov dimensionality dependence diagram d_L (Fig. 4.24(d)) calculated with the help of Kaplan–Yorke formula [60] prove that although the ε_1 parameter leads to chaotization of oscillations, it has no significant influence on the generated oscillation properties and leaves them quite homogeneous within a wide range of the ε_1 parameter change. Control of the generated oscillations properties, including those of chaotic oscillations, can be carried out with the help of coupling parameters due to the fact that generation of quasisynchronous oscillations at the output of individual generators is matched by a great diversity of attractors. As an example of a CMO with a set of various properties, Fig. 4.25 shows projections of the chaotic attractor model (4.2) with rotation indices equal to $[0, 0, 0]$, $[0, 0, 1]$, $[0, 1, 1]$, as well as their respective power spectra, self-correlation functions and Lyapunov d_L dimensionalities values calculated with the help of the Kaplan–Yorke formula. Here, oscillation properties at the first

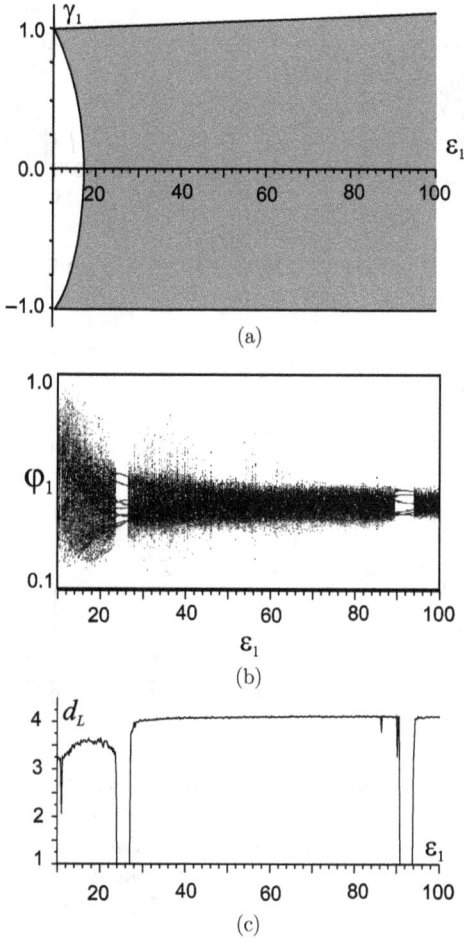

Fig. 4.24 (a) CMO existence region at the first CPLL generator output at $\gamma_2 = 0.5$, $\varepsilon_2 = 7.5$, $\gamma_3 = 0.7$, $\varepsilon_3 = 3$, $\kappa_1 = 0.15$, $\kappa_2 = 1.6$; (b) single-parameter bifurcation diagram of $\{\varepsilon_1, \varphi_1\}$ Poincaré map and (c) dependence of Lyapunov dimensionality of the d_L attractor responsible for the map formation for $\gamma_1 = 0.5$.

generator's output were changed through a variation of the single κ_2 parameter.

Two-parametrical maps of dynamic modes and diagrams of modulating oscillations properties shown in Figs. 4.26 and 4.27 give a more complete idea about chaotic modes of an ensemble with inertia control circuits. Figure 4.26 shows maps of the dynamic mode model

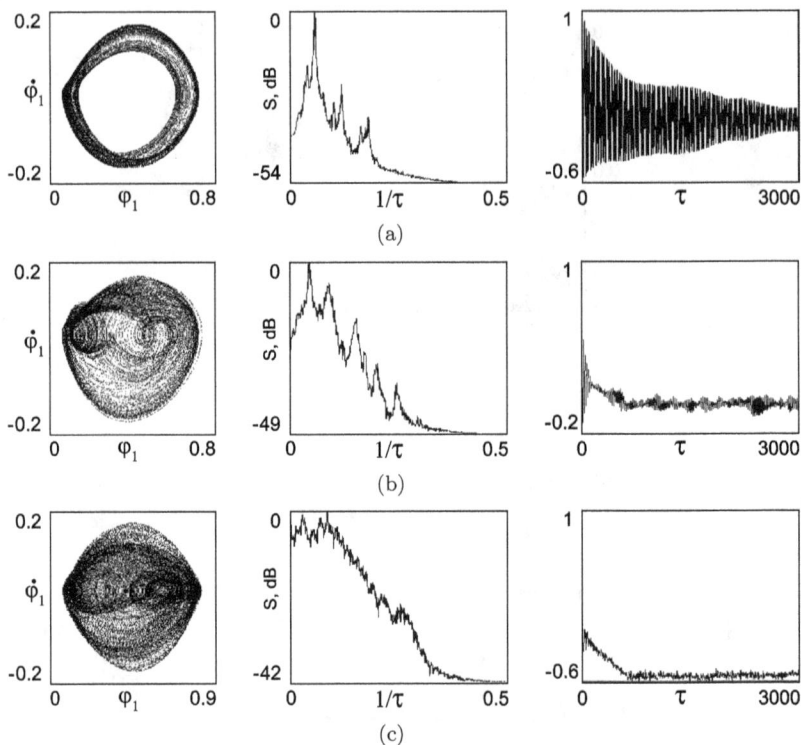

Fig. 4.25 Projections of the chaotic attractor model (4.2) on the partial subregion $U_1 = \{\varphi_1(\mathrm{mod}\, 2\pi), y_1\}$ and respective power spectrum and self-correlation functions calculated based on the $y_1(\tau)$ realization at (a) $\gamma_1 = 0.4$, $\varepsilon_1 = 0.8$, $\gamma_2 = 0.5$, $\varepsilon_2 = 7.5$, $\gamma_3 = 0.7$, $\varepsilon_3 = 3$, $\kappa_1 = 0.3$, $\kappa_2 = 0.512$ — attractor of the $[0,0,0]$ type; (b) $\kappa_2 = 0.84$ — $[0,0,1]$; (c) $\kappa_2 = 1.32$ — $[0,1,1]$.

(4.2) on the (κ_2, γ_1) and $(\varepsilon_1, \gamma_1)$ planes. From Fig. 4.26, we can see that a three-PLL ensemble with the first-order filters has a great diversity of dynamic modes, both regular and chaotic, and that existence regions of the said modes are unevenly distributed around the parameter space. Overall dimensions of the CMO pull-out regions for PLL$_1$, PLL$_2$ and PLL$_3$ in Fig. 4.26 are, respectively, equal to $D_{u1}^c = 0.63$, $D_{u2}^c = 0.13$ and $D_{u3}^c = 0.02$ in Fig. 4.26 are, respectively, equal to $D_{u1}^c = 0.79$, $D_{u2}^c = 0.56$ and $D_{u3}^c = 0.0$. Analysis of the dynamic mode maps enables us to identify on the one hand, parameters, whose infinitesimal variations can lead to significant changes in

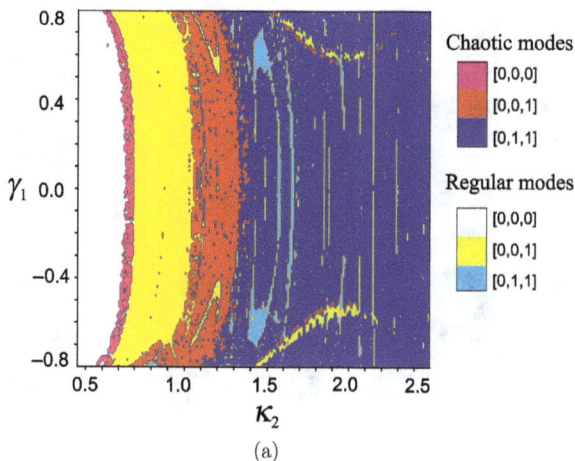

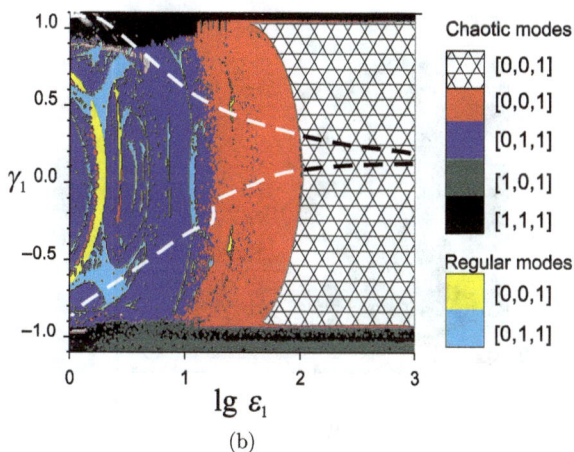

Fig. 4.26 Maps of dynamic modes of model (4.2) for (a) $\varepsilon_1 = 0.8$, $\gamma_2 = 0.5$, $\varepsilon_2 = 7.5$, $\gamma_3 = 0.7$, $\varepsilon_3 = 3$, $\kappa_1 = 0.15$ and (b) $\gamma_2 = 0.5$, $\varepsilon_2 = 7.5$, $\gamma_3 = 0.7$, $\varepsilon_3 = 3$, $\kappa_1 = 0.15$, $\kappa_2 = 1.6$.

the oscillations at the ensemble generator outputs, and on the other hand, parameters, whose variations cause less significant changes in the collective ensemble dynamics. Parameters, which have a pronounced effect, include additional coupling parameters κ_1 and κ_2. Partial PLL system parameters (initial frequency detuning γ_i and local control circuit inertia $\varepsilon_i, i = 1, 2, 3$) can be modified within a

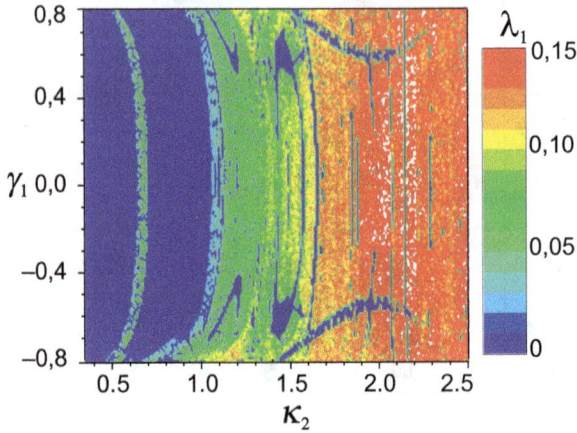

(a)

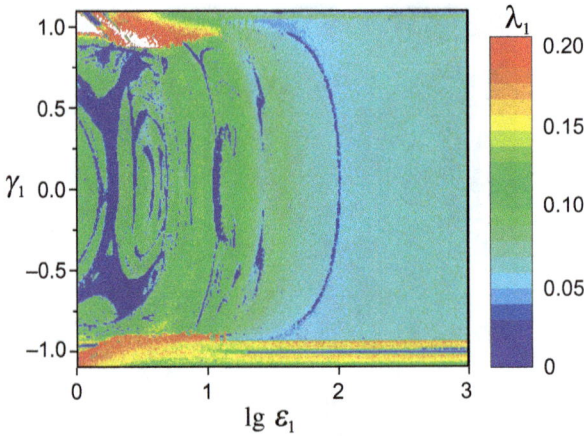

(b)

Fig. 4.27   Maps of Lyapunov maximum exponent of model (4.2) for (a) $\varepsilon_1 = 0.8$, $\gamma_2 = 0.5$, $\varepsilon_2 = 7.5$, $\gamma_3 = 0.7$, $\varepsilon_3 = 3$, $\kappa_1 = 0.15$ and (b) $\gamma_2 = 0.5$, $\varepsilon_2 = 7.5$, $\gamma_3 = 0.7$, $\varepsilon_3 = 3$, $\kappa_1 = 0.15$, $\kappa_2 = 1.6$.

relatively wide range as they have a less pronounced effect on the character of the oscillations produced by ensemble generators. The above can be clearly seen in the diagrams shown in Fig. 4.26. Chaotic homogeneity exponents of the D_{u1}^* and D_{u2}^* regions in Fig. 4.26, which are equal to $I_1^3 = 0.91$ and $I_2^3 = 0.9$, respectively, also prove the correctness of the above-stated facts.

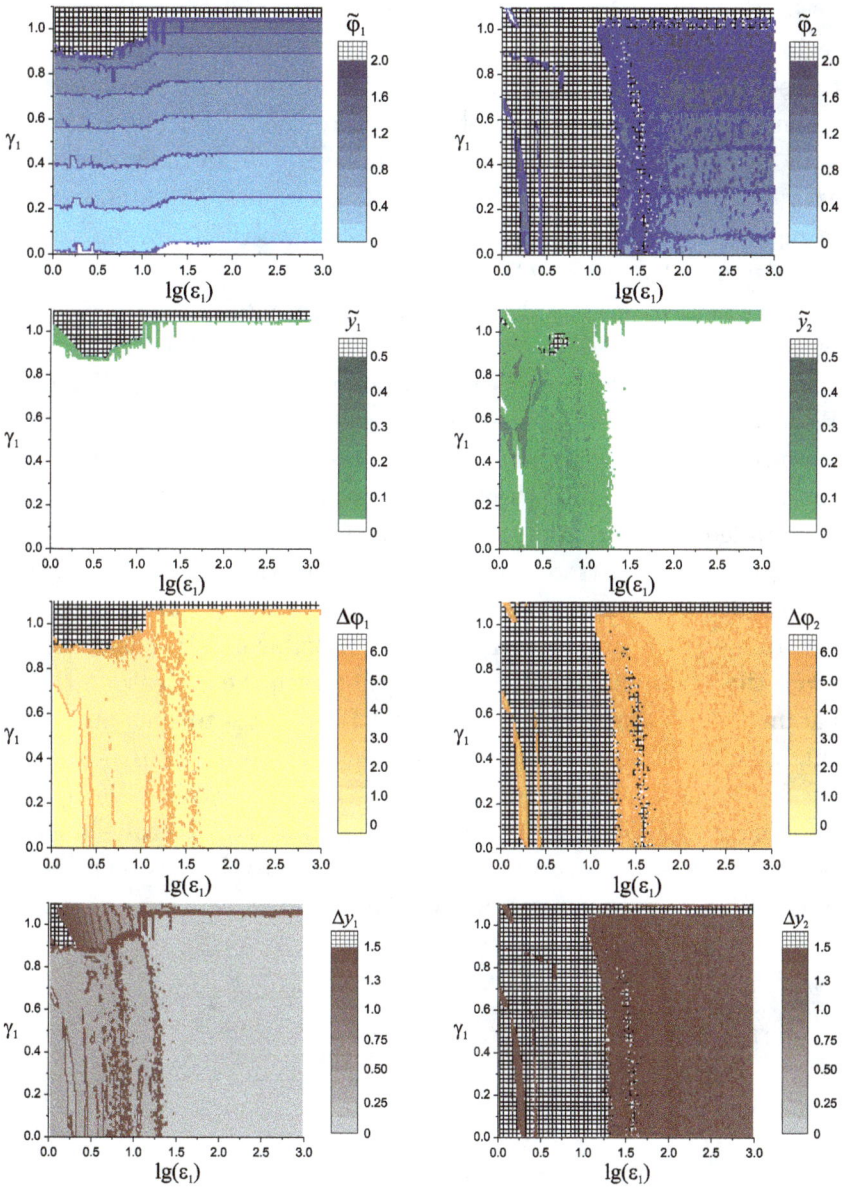

Fig. 4.28   Characteristics of modulating oscillations.

In Fig. 4.26, the dotted line outlines the D_{z1}^c and D_{z2}^c pull-in regions for the CMO modes of the first and second generators of the three-PLL ensemble under consideration. Like in a two-component ensemble, the three-PLL ensemble D_{z1}^c region practically coincides with the synchronization mode pull-in region of an individual PLL with an integrating filter in the control circuit. CMO-mode pull-in regions for a three-generator ensemble have the following overall dimensions: $D_{z1}^c = 0.29, D_{z2}^c = 0.1, D_{z3}^c = 0$.

Diagrams in Figs. 4.27 and 4.28 depict modulating oscillation properties. Figure 4.27 shows maps of Lyapunov maximum exponent, which characterize the degree of dynamic mode's chaotic character. We can see that the degree of dynamic mode's chaotic character increases following the increase in the κ_2 parameter caused by the variation of the quasisynchronism index and that it is preserved in a wide range of the γ_1 and ε_1 parameter variation.

Figure 4.28 illustrates the distribution of the $\tilde{\varphi}_{1,2}$ and $\tilde{y}_{1,2}$ coordinate mean values, as well as the $\Delta\varphi_{1,2}$ and $\Delta y_{1,2}$ coordinate variation range in the first two PLLs of the ensemble under consideration. From the above diagrams, it follows that with the increase in the ε_1 parameter the Δy_1 value in the first PLL reduces to such an extent that this self-modulation mode begins to rather correspond to the synchronization mode than to the CMO mode. In Fig. 4.26(b), the parameter region, where $\Delta\varphi_1 < 0.1$ and $\Delta y_1 < 0.02$, is cross-hatched. It is worth referring this cross-hatched section to the existence region of the PLL$_1$ synchronous mode.

Chapter 5

Phase System Parallel Coupling

5.1. Models and Modes

The structural scheme of an ensemble of two phase-locked loops connected in parallel (PPLL) is shown in Fig. 5.1. Here, the reference signal for both voltage controlled oscillators of the ensemble is signal $S_0(t)$. It comes to the phase discriminators PD_1 and PD_2, where it is compared with signals $S_1(t)$ and $S_2(t)$ of the voltage controlled oscillators G_1 and G_2, respectively. At the outlet of PD_1 and PD_2 voltages u_1 and u_2 are produced, which are proportional to the difference of oscillation phases of the voltage controlled oscillator and the reference signal. Further, the phase detuning signals u_1 and u_2 come to the summators Σ, where it is summarized with the signals of the couplings $u'_1 = \kappa u_2$ and $u'_2 = \delta u_1$. Signals $\tilde{u}_1 = u_1 + \kappa u_2$ and $\tilde{u}_2 = u_2 + \delta u_1$ are used for managing the frequencies of generators G_1 and G_2 with control circuits CC_1 and CC_2. Operation equations describing such system dynamics are as follows [4]:

$$\frac{p\varphi_1}{\Omega_1} = \frac{\Omega_1^0}{\Omega_1} - K_1(p)[F(\varphi_1) + \kappa F(\varphi_2)],$$

$$\frac{p\varphi_2}{\Omega_1} = \frac{\Omega_2^0}{\Omega_1} - bK_2(p)[F(\varphi_2) + \delta F(\varphi_1)],$$

(5.1)

where $p \equiv d/dt$, φ_i is the current phase detuning, Ω_i^0 is the initial frequency detuning of the ith controlled generator relative to the reference signal, $b = \Omega_2/\Omega_1$, Ω_i is the pull-out range of PPLi, $K_i(p)$ is the low frequency filter F_i transmission coefficient, $F(\varphi_i)$

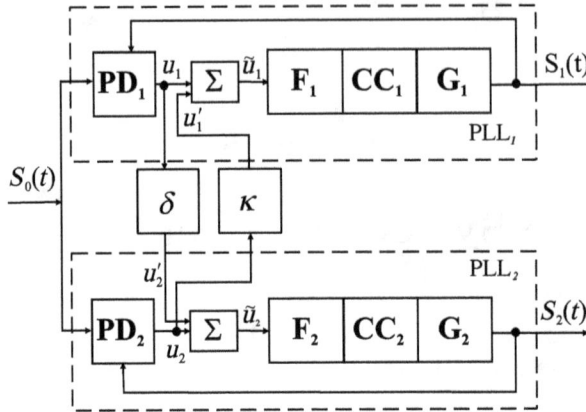

Fig. 5.1 Structural scheme of the parallel coupling of two PLLs.

the standardized feature of the ith phase discriminator $(i = 1, 2)$, κ and δ are coupling parameters.

We assume similar pull-out ranges of partial systems $\Omega_1 = \Omega_2 = \Omega$, and sinusoidal characteristics of the phase discriminators $F(\varphi_{1,2}) = \sin(\varphi_{1,2})$. By using the second-order filters in control circuits with transmission coefficients $K_1(p) = (1 + a_1 p + b_1 p^2)^{-1}$ and $K_2(p) = (1 + a_2 p + b_2 p^2)^{-1}$, from Eq. (5.1), the following differential equation system can be obtained:

$$\frac{d\varphi_1}{d\tau} = y_1, \quad \frac{dy_1}{d\tau} = z_1,$$

$$\mu_1 \frac{dz_1}{d\tau} = \gamma_1 - y_1 - \varepsilon_1 z_1 - \sin\varphi_1 - \kappa \sin\varphi_2$$

$$\equiv P(\varphi_1, y_1, z_1, \varphi_2, y_2, z_2), \qquad (5.2)$$

$$\frac{d\varphi_2}{d\tau} = y_2, \quad \frac{dy_2}{d\tau} = z_2,$$

$$\mu_2 \frac{dz_2}{d\tau} = \gamma_2 - y_2 - \varepsilon_2 z_2 - \sin\varphi_2 - \delta \sin\varphi_1$$

$$\equiv Q(\varphi_1, y_1, z_1, \varphi_2, y_2, z_2).$$

The system is defined in the six-dimensional cylindrical phase space $U = \{\varphi_1 (\mathrm{mod}\ 2\pi), y_1, z_1, \varphi_2\ (\mathrm{mod}\ 2\pi), y_2, z_2\}$. Here, $\tau = \Omega t$ is the non-dimensional time, $\varepsilon_i = \Omega a_i > 0$, $\mu_i = \Omega^2 b_i > 0$ is the non-dimensional filter parameters, $\gamma_i = \Omega_i^0/\Omega$ the relative initial frequency detuning of the controlled generators, $i = 1, 2$.

At $\mu_1 \ll 1$, $\mu_2 \ll 1$, the system (5.2) has small parameters at derivatives dz_1/dt and dz_2/dt. General motion in the phase space U is divided into "rapid" and "slow" motions. The surface $W : \{P(\varphi_1, y_1, z_1, \varphi_2, y_2, z_2) = 0, \ Q(\varphi_1, y_1, z_1, \varphi_2, y_2, z_2) = 0\}$ of "slow" motion is stable for rapid motion. Slow motion equations on the surface W are as follows:

$$\frac{d\varphi_1}{d\tau} = y_1,$$

$$\varepsilon_1 \frac{dy_1}{d\tau} = \gamma_1 - \sin \varphi_1 - y_1 - \kappa \sin \varphi_2 \equiv P_1(\varphi_1, y_1, \varphi_2, y_2),$$

$$\frac{d\varphi_2}{d\tau} = y_2, \tag{5.3}$$

$$\varepsilon_2 \frac{dy_2}{d\tau} = \gamma_2 - \sin \varphi_2 - y_2 - \delta \sin \varphi_1 \equiv Q_1(\varphi_1, y_1, \varphi_2, y_2).$$

The system (5.3) is defined in the four-dimensional cylindrical phase space $V = \{\varphi_1(\text{mod } 2\pi), \ y_1, \ \varphi_2 \,(\text{mod } 2\pi), \ y_2\}$. It is a mathematical model of PPLL with integrating filters $(K_{1,2}(p) = (1 + a_{1,2}p)^{-1})$ in the control circuits, where $\varepsilon_{1,2} = \Omega a_{1,2}$.

At $\varepsilon_1 \ll 1$, $\varepsilon_2 \ll 1$, the system (5.3) has a stable integral surface $W_1 : \{P_1(\varphi_1, y_1, \varphi_2, y_2) = 0, \ Q_1(\varphi_1, y_1, \varphi_2, y_2) = 0\}$, on which motions are defined by the system of equations:

$$\frac{d\varphi_1}{d\tau} = \gamma_1 - \sin \varphi_1 - \kappa \sin \varphi_2,$$

$$\frac{d\varphi_2}{d\tau} = \gamma_2 - \sin \varphi_2 - \delta \sin \varphi_1. \tag{5.4}$$

The periodicity of variables φ_1 and φ_2 on the right-hand side of the system (5.4) indicates that it is a nonlinear dynamic system on the toroidal phase surface $T = \{\varphi_1(\text{mod } 2\pi), \varphi_2(\text{mod } 2\pi)\}$. It describes the dynamics of PPLL with idealized filters $(K_{1,2} = 1)$ in the control circuits.

As mathematical models of ensembles in parallel and cascade couplings of phase-locked loops (PLL) are defined in the same phase spaces, the earlier completed correspondence for the cascade coupling of PLL between the attractors of mathematical models and the dynamic modes of the controlled generators at the parallel coupling is kept in force. While studying the ensemble's dynamic features, the approach connected with continuation of the structures placed in the parameter space along the parameter is examined. At first, dynamics of the model (5.4), PPLL without filters in the control circuits is examined, where the role of coupling parameters is studied, then the set of structures continues along the parameters $\varepsilon_1, \varepsilon_2$ in the model (5.3), and μ_1, μ_2 in the model (5.2), where the role of parameters of the control circuits is studied [67].

5.2.   Ensemble with Low-Inertia Control Circuits

In this section, asynchronous mode PPLL with the first-order filters is examined, when $\varepsilon_1 \ll 1$ and $\varepsilon_2 \ll 1$. In this case, the ensemble behavior can be described through the motion of the model (5.4). Examination is performed for strong and weak couplings, which helps to identify the role of couplings in the ensemble dynamics. The couplings with the coupling coefficients κ and δ less than unity will be referred to as the weak ones, and the ones with the coefficient more than unity will be referred to as the strong ones.

5.2.1.   *Weak coupling case*

Figure 5.2(a) shows the division of the plane (γ_1, γ_2) of initial returning frequency of the model (5.4) in the case when both couplings are weak $\delta = 0.5, \kappa = 0.7$. Figure 5.2(b) depicts the structure of the space parameters (5.4) in case when one of the couplings is strong ($\kappa = 1.8 > 1$), the other is weak ($\delta = 0.2 < 1$), and the totality of the couplings remains weak ($\delta\kappa < 1$). In Fig. 5.2, the region C_0 of the equilibrium state is a parallelogram:

$$C_0 = \left\{ \max\left[\frac{(\kappa\delta - 1) \cdot z \cdot \operatorname{sign}\delta + \gamma_2}{\delta}, (\kappa\delta - 1) \cdot z + \kappa\gamma_2 \right] \right.$$

(a)

(b)

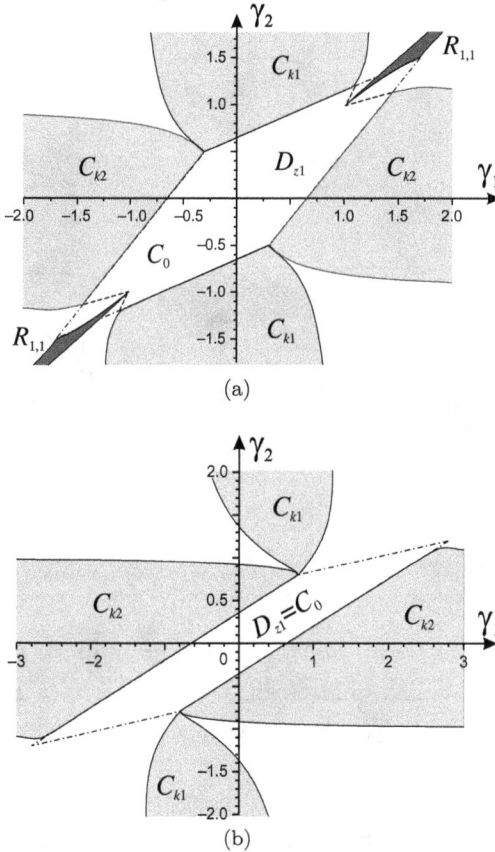

Fig. 5.2   The structure of division of the parameter space of PPLL with low-inertia control circuits for weak couplings (a) $\delta = 0.5, \kappa = 0.7$ and (b) $\delta = 0.2$, $\kappa = 1.8$.

$$< \gamma_1 < \min\left[\frac{(1 - \kappa\delta) \cdot z \cdot \mathrm{sign}\,\delta + \gamma_2}{\delta}, (1 - \kappa\delta) \cdot z + \kappa\gamma_2\right]\right\},$$

$$z = \mathrm{sign}(1 - \kappa\delta). \tag{5.5}$$

At parameter values from the region C_0 in the phase space of the model (5.4), the stable equilibrium state O_1 always exists (Fig. 5.3(a)), which defines the synchronous mode I_{S1}. That is why the region C_{S1} of existence (keeping) of the mode I_{S1} coincides with the region C_0. Leaving C_{S1} for the region C_{k2}, the equilibrium states

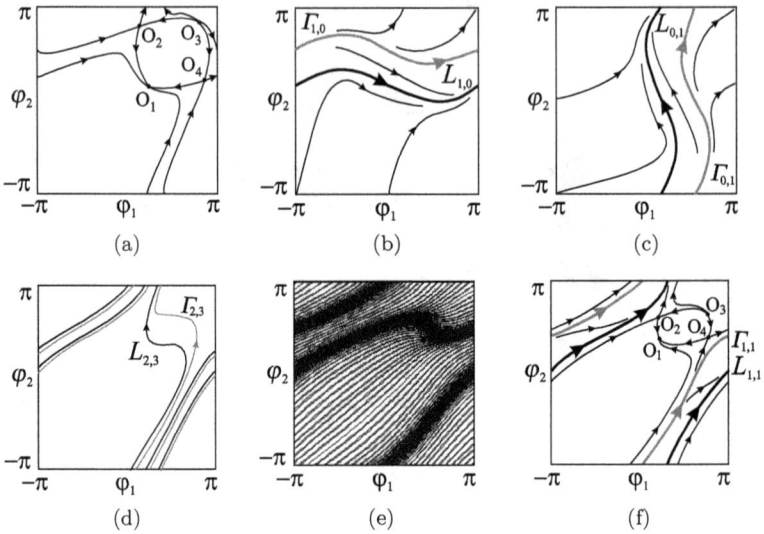

Fig. 5.3   Phase portraits of the model (5.4) for weak couplings.

disappear; in the phase space (5.4) a pair of oscillatory–rotatory limit cycles is created, which encircles the phase torus T in the direction φ_1 (Fig. 5.3(b)) i.e., O_1. The stable cycle $L_{1,0}$ defines the mode in which the first generator functions in the beat mode and the second generator functions in the quasisynchronism mode; in PPLL, the partial quasisynchronization mode I_{k2} is installed. The boundaries of the region C_{k2} are the bifurcation curves of double limit cycles and saddle-node separatrix loops which encircle the phase torus. Leaving C_{S1} for the region C_{k1}, the disappearance of equilibrium states is accompanied by creation of oscillatory–rotatory limit cycles which encircle the phase torus T in the direction φ_2 (Fig. 5.3(c)). The stable cycle $L_{1,0}$ corresponds to the fact that the second generator functions in the beat mode and the first generator functions in the quasisynchronism mode; in PPLL, the partial quasisynchronization mode I_{k1} is installed. The boundaries of the regions C_{k1} are bifurcation curves of double limit cycles and saddle-node separatrix loops. Outside the regions C_{S1}, C_{k1} and C_{k2} in the phase space of the model (5.4) there are motions encircling the torus itself in the direction φ_1 as well as in the direction φ_2. These motions are characterized by the rotation number ν: if ν is a rational number and $\nu \neq 0, \nu \neq \infty$, then rotatory limit cycles respond to it (Fig. 5.3(d)); if ν is irrational number, then there is the torus T which is quasiwinding (Fig. 5.3(e)). All rotatory attractors define the global mode of beatings.

In Fig. 5.2(b) in the region C_0, the equilibrium state O_1 is globally stable, that is why C_0 is also the region D_{Z1} of pull-in to the synchronous mode I_{S1}. In the case when both couplings are weak (Fig. 5.2(a)), the global stability of the synchronous mode I_{S1} can be disturbed because of the stable rotatory cycle $L_{1,1}$ appearing in the phase space of the model (5.4) (Fig. 5.3(f)). The region $R_{1,1}$ in the cycle $L_{1,1}$ is marked dark in Fig. 5.2(a). It is shown in the region C_0, that is why the pull-in region $D_{Z1} = C_0 \backslash R_{1,1}$ in the mode I_{S1} is less than the region C_{S1}.

5.2.2. *Strong coupling case*

Phase portrait of the system (5.4) on the plane (γ_1, γ_2) for strong couplings is shown in Fig. 5.4. Here, the chain line limits the region C_0. The rest of the lines correspond to the following bifurcations:

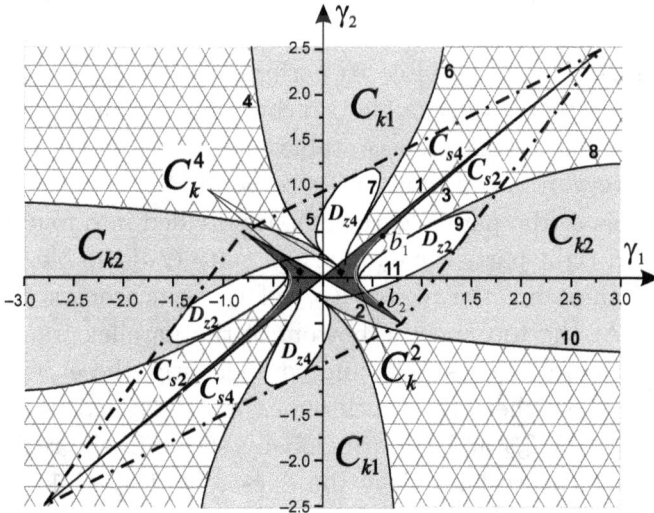

Fig. 5.4 The structure of division of the parameter space of PPLL with low-inertia control circuits for strong couplings $\delta = 1.5, \kappa = 1.8$.

lines 1 and 2 correspond to the changes in the equilibrium state stabilities O_2 and O_4; changes in stabilities O_4 and O_2 occur smoothly and not so smoothly, respectively;

line 3 limits the region of oscillatory limit cycle existence. It consists of the curve which corresponds to the saddle-node bifurcation (between points b_1 and b_2) and the curve which corresponds to separatrix loops, which do not encircle the phase torus T itself;

lines 4, 6 and 8, 10 correspond to the saddle-node bifurcations of oscillatory–rotatory limit cycles encircling torus T itself in the directions φ_2 and φ_1, respectively.

lines 5, 7 and 9, 11 respond to the separatrix loops encircling torus T itself in the directions φ_2 and φ_1, respectively.

As the system (5.4) is invariant relative to the change of $(\gamma_1, \gamma_2, \varphi_1, \varphi_2)$ into $(-\gamma_1, -\gamma_2, -\varphi_1, -\varphi_2)$, then the rest of the lines in Fig. 5.4 are not numbered, and they turn into the numbered lines for the above-mentioned transformation. Now, we will come to the analysis of the regions which were created by the given bifurcation curves, and dynamic modes realized in these regions. It is remarkable that the systems on the torus have denumerable number of different types of attractors. It is impossible to perform their complete examination. That is why, we will examine synchronous and quasisynchronous modes, which in the phase space stable equilibrium states, oscillatory and oscillatory–rotatory limit cycles respond to.

Diagonals of the parallelogram C_0 are divided into four parts, in the left and right parts the equilibrium state O_2 is stable, and consequently, these are the regions C_{S2} of the synchronous mode I_{S2} existence. At the top and the bottom of the parallelogram C_0, the equilibrium state O_4 is stable, and consequently, these parts create the region C_{S4}, where the synchronous mode I_{S4} exists. The synchronous modes I_{S2} and I_{S4} are globally stable in the regions D_{Z2} and D_{Z4}, which are limited by the lines 3, 9, 11 and 1, 5, 7. The regions D_{Z2} and D_{Z4} are the pull-in regions into the synchronous modes I_{S2} and I_{S4} respectively; the boundaries of these regions are Andronov–Hopf bifurcation curves and the separatrix loops which

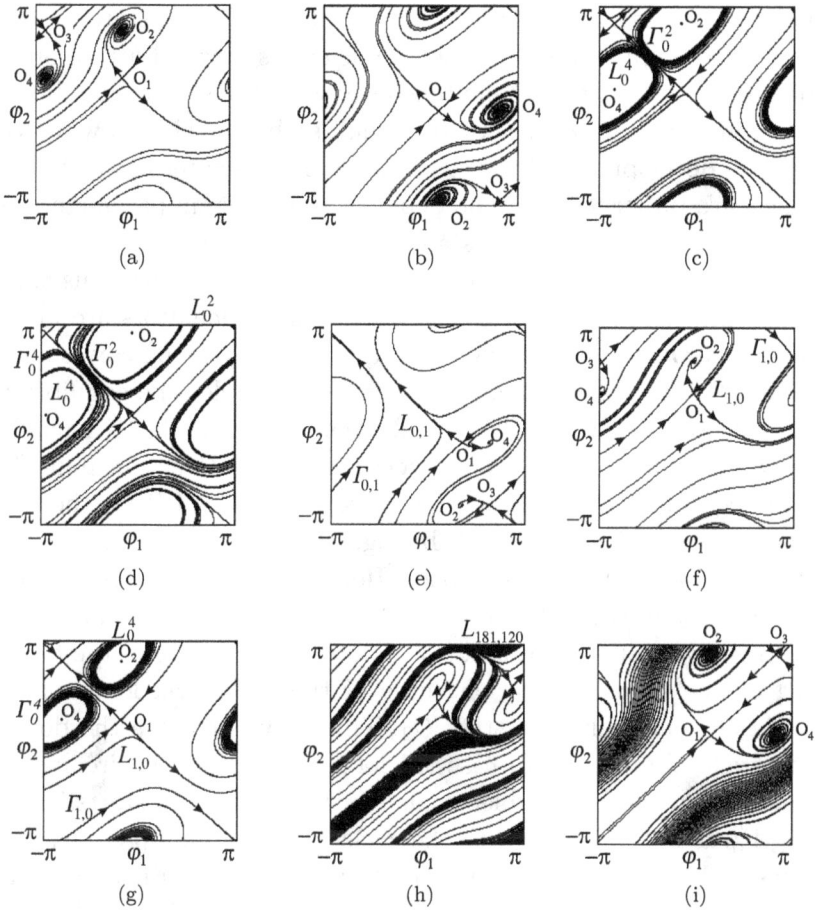

Fig. 5.5   Phase portraits of the model (5.4) for strong couplings.

encircle torus T itself. In Fig. 5.4 the pull-in regions into the synchronous mode are marked white. The phase portraits responding to the regions D_{Z2} and D_{Z4} are shown in Figs. 5.5(a) and 5.5(b), respectively. These portraits are the result of computer simulation, which is complemented with equilibrium state codes, and arrows show the direction of motion along the phase trajectories. Limit cycles on the phase portraits are characterized by multiplexing of the phase trajectories in their surroundings. It is remarkable that the region D_{Z4} consists of two parts which are divided by the narrow regions in the modes of beatings. The global stability of equilibrium states is broken with the appearance of stable limit cycles. Several attractors of different types can exist simultaneously with the stable equilibrium states on the torus T, i.e., there is multistability. That is why, scenarios of development of the ensemble dynamics entering the pull-in regions are rather variable and depend on the system parameters, as well as on the system initial state.

Lines 1 and 3 limit existence of the global quasisynchronous mode I_{k4} in the region C_k^4, in which both ensemble generators are in the quasisynchronous mode. The oscillatory limit cycle L_0^4 corresponds to the mode I_{k4} on the phase torus which encircles the equilibrium state O_4 (Fig. 5.5(c)). The unstable oscillatory cycle Γ_0^2 encircles the equilibrium state O_2. This cycle limits the attraction pool Π_{02} of the equilibrium state O_2; at $(\varphi_1, \varphi_2) \in \Pi_{02}$ in PPLL, the synchronous mode I_{S2} is realized. In Fig. 5.4, the region C_k^4 is marked dark, the boundaries are Andronov–Hopf bifurcation curves, the separatrix loops of the saddle O_1, encircling O_4, are of the double limit cycles (between points b_1 and b_2). Points b_1 and b_2 are at the beginning of one more bifurcation curve of the separatrix loops of the saddle O_1, encircling the state O_2. This curve precedes the double limit cycle bifurcation. In case the loop bifurcations, encircling O_2, are on the torus T a stable oscillatory limit cycle L_0^2 starts, encircling Γ_0^2. Simultaneously with the cycle L_0^2 from the separatrix loop of the saddle O_3, the unstable limit cycle Γ_{04} starts encircling L_0^4 (Fig. 5.5(d)). So, the double limit cycle and separatrix loop bifurcation curves are singled out on the parameter plane of the region C_k^2 at parameter values from which L_0^2 is derived on the torus T. In Fig. 5.4 the region C_k^2 is marked lighter than C_k^4. The cycle L_0^2 defines the global quasisynchronization mode I_{k2} which differs from the mode I_{k4} with average values $\tilde{\varphi}_1$, $\tilde{\varphi}_2$ and deviations $\Delta\varphi_1$, $\Delta\varphi_2$. As C_k^2 is crossed by C_k^4, then this region is the region of a multistable behavior. Here, depending on initial terms, it is possible to install modes I_{S2}, I_{k2}, I_{k4} (Fig. 5.5(d)).

Regions C_{k1} and C_{k2} are the regions of existence of partial quasisynchronization modes I_{k1} and I_{k2}; on the torus T the stable oscillatory–rotatory cycles $L_{0,1}$ and $L_{1,0}$ correspond to them accordingly. In Fig. 5.4, these regions are marked light gray, their boundaries are separatrix loops and the double limit cycles encircling torus T itself are the bifurcation curves. Outside the region C_0, modes I_{k1} and I_{k2} are globally stable (phase portraits are equivalent to portraits in Figs. 5.3(b) and 5.3(c)), in the regions of crossing with C_0 they always exist together with synchronous modes (Figs. 5.5(e) and 5.5(f)). Besides, there are parameter values where they are realized together with global quasisynchronization modes (Fig. 5.5(g)).

In the regions marked with shading on the torus T, rotatory attractors can exist, which are responsible for the installation of global modes of beatings. Outside the region C_0, they are either rotation limit cycles or quasiperiodic trajectories (phase portraits are equivalent to portraits in Figs. 5.3(d) and 5.3(e)). In the region C_0, only rotation limit cycles can exist but they can be of different complexity — from simple, containing one rotation by 2π along the coordinates φ_1 and φ_2, to complex, including numerous rotations along φ_1 and φ_2. For numerous rotations, the number of which, as a rule, at variables φ_1 and φ_2 is not the same, complex limit cycles resemble chaotic attractors more than the regular ones. In Fig 5.5(h), as an example, a type of a stable limit cycle with the rotation $\nu = 181/120$ is shown. Regions of modes of beatings are created in the C_0 region's layer structure, similar to the structure of the parameter space found in a cascade type of coupling — while moving in the shading region C_0, regions of the global stability of synchronous modes alternate with regions of modes of beatings. In this part of the parameter space, transient periods to globally stable synchronous modes can be rather long and can have a complex sight (see Fig. 5.5(i)). Here, the equilibrium states O_4 and O_2 are stable and unstable, respectively. If the system initial state is situated in the surrounding O_2, then transfer to O_4 will be long and will have a complex sight.

So, in the framework of the dynamic models (5.3) and (5.4), examination of dynamic modes of the parallel coupling of two PLL systems with low-inertia control circuits was performed. Computer study helped us to define the role of couplings in the parallel coupling of the systems into an ensemble.

It is established that coupling of phase systems with weak couplings leads to regular quasisynchronous modes which are not relevant for partial systems. Quasisynchronous mode regions are not crossed with synchronous modes and beating mode regions, that is why quasisynchronous modes are always globally stable. In case when both couplings are weak ($\kappa < 1, \delta < 1$), the region of pull-in to a synchronous mode is less than the region of pull-out. At the expense of changing the coupling force, which remains in the framework of a total weak coupling ($\kappa < 1, \delta < 1$), it is possible to obtain coincidence of the regions of pull-in and pull-out of a synchronous mode. Global quasisynchronous modes, when both generators function in a quasisynchronous mode, in the PPLL with weak couplings are not established.

The ensemble of phase systems with strong couplings has a set of new effects which are not peculiar to the ensemble with weak couplings. With strong couplings in the ensemble, global quasisynchronous modes appear. These modes always exist inside the region C_0, that is why, firstly, they have limit dimensions, secondly, these modes cannot be globally stable. Regions of partial quasisynchronization modes and modes of beatings enter the region of synchronous modes, and this leads to the appearance of the layer structure of the parameter space, multistability and hysteresis phenomena. The regions of pull-in to synchronous modes do not coincide with the regions of pull-out; the regions of pull-in to a synchronous mode consist of several subregions.

Comparative analysis of the research results of cascade and parallel types of coupling shows that a lot of phenomena studied before a cascade coupling hold, e.g., the existence of several synchronous modes, appearance of global and partial quasisynchronization modes, bifurcation mechanisms of the appearance of quasisynchronous modes, the existence of layer structure of the parameter space. Some differences are observed in correlation to system parameters and couplings, when these or the other modes are realized. Particularly, with parallel coupling, global synchronization modes appear in the region of little frequency detuning, at weak couplings, quasisynchronous modes are always globally stable, the regions of pull-in to a synchronous mode consist of several subregions.

5.3. Filters Parameters Influence

5.3.1. *The first-order filter case*

We will study the role of filters in the control circuits in the dynamics of the examined ensemble with the case when a filter is only in one of the unified subsystems. Let the first-order filter be in the second subsystem. The changes taking place in the parameter space of the model (5.3) at increasing time constant of the filter ε_2 are illustrated by the diagrams in Fig. 5.6.

Figure 5.6(a) corresponds to a comparatively small time constant $\varepsilon_2 = 1$. Comparing this portrait to the analogous one in Fig. 5.2(a) for $\varepsilon_2 = 0$, we should note the changes, which occur in the ensemble as the filter time constant grows. Firstly, the existence of regions of quasisynchronous oscillations increases, secondly, the region C_{k1} at $\gamma_2 > 0$ moves to the left, and the region C_{k2} at $\gamma_1 < 0$ moves to the top. For this movement, regions C_{k1} and C_{k2} start to overlap, that is, the global stability of quasisynchronous modes is broken. The changes also concern synchronous and asynchronous modes. The region $R_{1,1}$ with the existence of the rotation limit cycle $L_{1,1}$ becomes larger, as well as intersection of the regions $R_{1,1}$ and C_0 increases, that leads to decrease in the region D_{Z1} of pull-in to the synchronous mode I_{S1}. New scenarios of the ensemble entering into a synchronous mode appear. A double period of the cycle $L_{1,1}$ bifurcation occur at the exit of the region $R_{1,1}$ in the area (a_1, a_2). As a result, the chaotic rotatory attractor, which defines the chaotic mode of beatings, is formed. When decreases further γ_1 (increasing γ_2), a chaotic attractor breaking occurs, and all phase trajectories of the model (5.3) tend to be of the equilibrium state O_1. So, pull-in to a synchronous mode is accompanied by chaotization of the mode of beatings.

In Fig. 5.6(b), a two-parameter bifurcation diagram of the model (5.3) at $\varepsilon_2 = 10$ is shown. The regions D_{k1} and D_{k2} of pull-in to the quasisynchronous modes I_{k1} and I_{k2} are marked dark

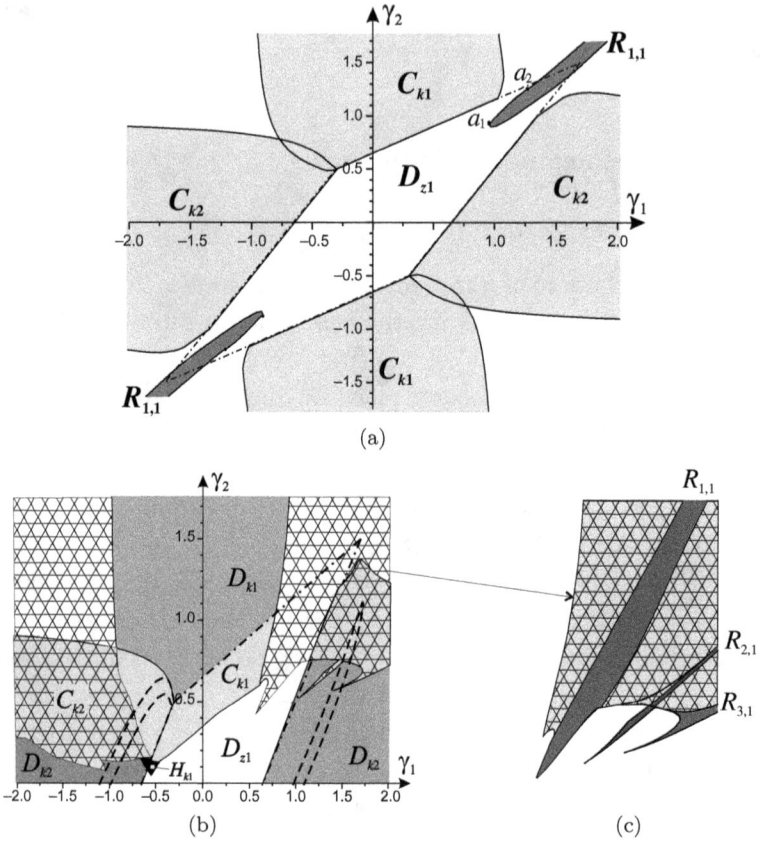

Fig. 5.6 The structure of division of the parameters space of the model (5.3) at $\kappa = 0.7$, $\delta = 0.5$, (a) $\varepsilon_1 = 0$, $\varepsilon_2 = 1$ and (b) $\varepsilon_2 = 10$, (c) a fragment of the asynchronous motions existence region.

correspondingly, the regions C_{k1} and C_{k2}, where these modes coexist, are marked light. The region H_{k1}, marked black in the figure, is the region where chaotic oscillations exist with the rotation index $[0, 1]$ — the region of chaotic-modulated oscillations at the outlet of the first ensemble generator. At the examined parameters, the region of existence of the synchronous mode I_{S1} coincides with the region C_0, the boundaries of which are marked with a dash-and-dot line. The region D_{Z1} of pull-in to the mode I_{S1} in Fig. 5.6(b) is marked white. The dotted lines in Fig. 5.6(b) correspond to the double period of the cycle $L_{1,0}$ bifurcation, which is the image of the mode I_{k2}. The area of the beating mode is marked with shading. In contrast to the case $\varepsilon_2 = 1$, here, the space of asynchronous modes, which can break the global stability of a synchronous and quasisynchronous modes, is constructed in a rather complicated way. It includes the regions of existence of rotatory attractors of different types. Figure 5.6(c) shows the structure of asynchronous motions, it is a fragment of Fig. 5.6(b), containing parameter regions of asynchronous modes. In Fig. 5.6(c), some regions composing it are marked (in Fig. 5.7(b) they are not marked): $R_{1,1}$, $R_{2,1}$ and $R_{3,1}$ are the regions of existence of cycles $L_{1,1}$ (Fig. 5.7(a)), $L_{2,1}$ (Fig. 5.7(b)) and $L_{3,1}$ (Fig. 5.7(c)), respectively.

The regions of existence of different types of attractors can overlap, which amplifies the ensemble's of multistable features. The boundaries of the singled out regions are bifurcation curves of limit cycles generating (saddle-node bifurcations and separatrix loops bifurcations) and their changing stability. Cycles can lose their stability in case of coming of a pair of the complex-conjugate multipliers

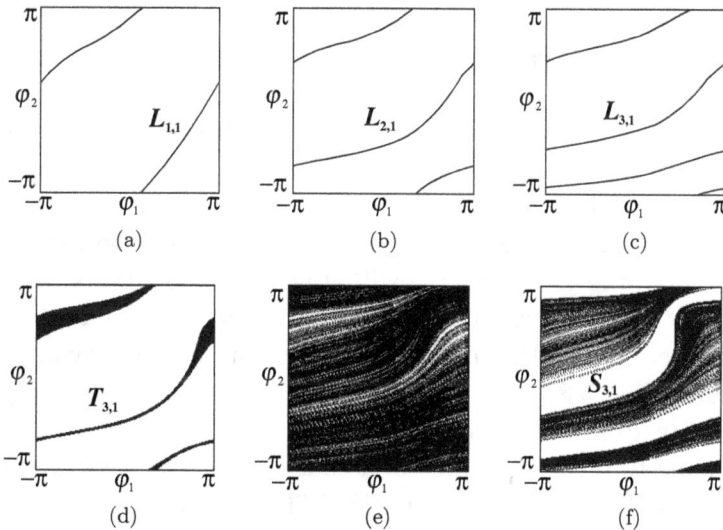

Fig. 5.7   The projections of rotatory attractors of the model (5.3) at $\kappa = 0.7$, $\delta = 0.5$, $\varepsilon_1 = 0$, $\varepsilon_2 = 10$, (a) $\gamma_1 = 1.29$, $\gamma_2 = 1.26$; (b) $\gamma_1 = 1.35$, $\gamma_2 = 0.6949$; (c) $\gamma_1 = 1.64$, $\gamma_2 = 0.6381$; (d) $\gamma_1 = 1.35$, $\gamma_2 = 0.705$; (e) $\gamma_1 = 1.944$, $\gamma_2 = 0.692$; (f) $\gamma_1 = 1.684$, $\gamma_2 = 0.6839$.

onto a unit circle as in the case of a double period bifurcation. Both these bifurcations can be either soft or hard.

The complex structure of the boundaries of the regions of beatings creates a great variety of bifurcation transfers at parameter variations of the model (5.3). It is established, for example, that leaving the regions $R_{1,1}$, $R_{2,1}$, $R_{3,1}$ depending on a leaving direction, there is a possibility of appearance of more complicated modes of beatings, including two-frequency and chaotic modes of partial quasisynchronization, as well as the installation of a global synchronization mode. The examples of rotatory attractors of different complexities are shown in Fig. 5.7. The invariant torus $T_{2,1}$ (Fig. 5.7(d)) defines the two-frequency mode of beatings. It appears from leaving of a pair of the unit multipliers of the cycle $L_{2,1}$ onto the unit circle. The attractor shown in Fig. 5.7(e), is also responsible for the two-frequency mode of beatings. This torus appears from a saddle-node bifurcation. It should be noted that invariant tori despite being regular attractors have a rather complicated structure. That is why their projections differ little from chaotic attractor projections. In the process of computer simulation, it is established that chaotic attractors of the model (5.3) can appear with different kinds of intermittency, in case of losing the smoothness of invariant tori and according to Feigenbaum scenario. Figure 5.7(f) shows a chaotic attractor projection, which, at variations of γ_1, appears through bifurcations of doubling the period of the cycle $L_{3,1}$, and at variations of γ_2 through the first kind of intermittency. As it can be seen from the represented projections, the model PPLL with the first-order filters has more variety of rotation-type attractors; that is why there is a great variety of scenarios of PPLL entering synchronous and quasisynchronous modes. In Fig. 5.8, there are diagrams of the processes accompanying pull-in to a quasisynchronous mode I_{k2}. According to them, it is possible to follow scenarios of a chaotic attractor generation, the projection of which is shown in Fig. 5.7(f).

The above-described phenomena of the complex dynamics of the ensemble with an inertia control circuit of PLL$_2$ also occur when a time lag effect is introduced into the control circuit of PLL$_1$. In this case, the phenomena of motion chaotization start with filters'

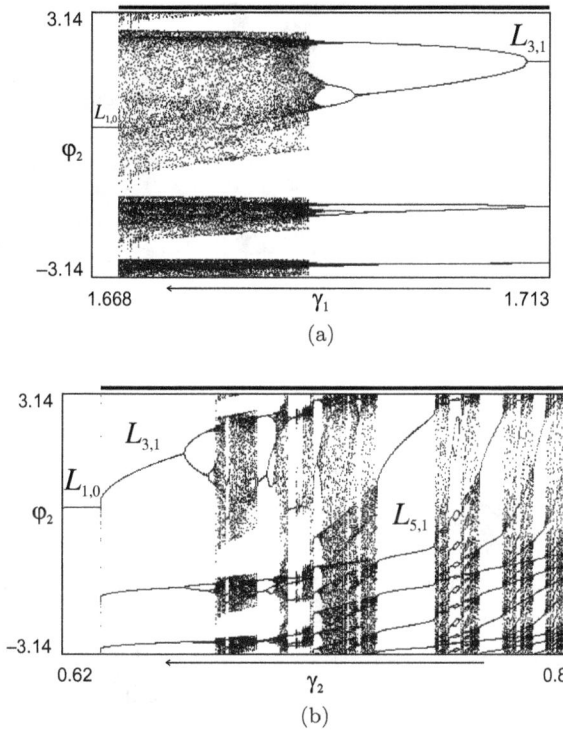

Fig. 5.8 One-parameter bifurcation diagram of Poincaré mapping of the model (5.3) at $\kappa = 0.7$, $\delta = 0.5$, $\varepsilon_1 = 0$, $\varepsilon_2 = 10$, (a) $\gamma_2 = 0.6839$ and (b) $\gamma_1 = 1.684$. The dark bend above the diagram characterizes rotation by the variable φ_2, the arrow shows the direction of changing of parameters γ_1 and γ_2.

parameter values which are smaller than in the case when a filter is in one of the subsystems. In Fig. 5.9, there is a one-parameter bifurcation diagram of Poincaré mapping illustrating the influence of the parameters of filters of control circuits on the ensemble dynamic modes. To create this diagram based on the initial system state (5.3), an oscillatory limit cycle L_0 was chosen, and it characterizes a global regular quasisynchronous mode. As the time constant of the filter ε_1 increases, the cycle L_0 through the cascade of the double period bifurcations turns into the chaotic attractor S_0 of the oscillatory type, which corresponds to the generation of chaotically modulated oscillation (CMO) at the outlets of both ensemble generators. As ε_1 grows, the attractor S_0 is transformed into a chaotic rotatory–oscillatory

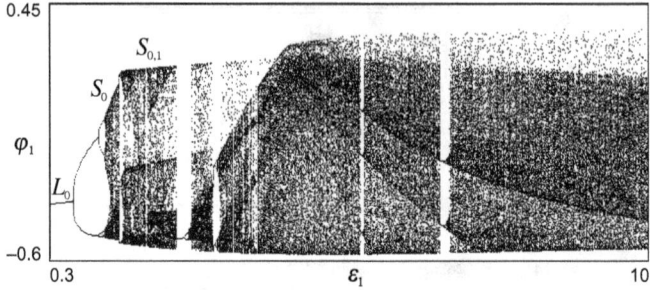

Fig. 5.9   Evolution of the oscillatory limit cycle L_0 of the model (5.3) at $\kappa = -0.4$, $\delta = 1.5$, $\gamma_1 = \gamma_2 = 0.3$, $\varepsilon_2 = 3$ with growing of ε_1.

attractor of the type $[0, 1]$. For this bifurcation, the second generator of the ensemble breaks away to the chaotic mode of beatings, the first one continues functioning in the mode of generation of CMO. Further increase in ε_1 does not qualitatively change the installed mode. The windows of regular motions on the represented diagram are narrow, and they are almost absent at $\varepsilon_1 > 10$.

It should be noted that dynamics of PLL with the first-order filters is regular and, consequently, the established phenomenon of the complex dynamics is a result of systems unification with non-linear couplings. The force and the structure of the couplings are crucial for the generation of chaotic oscillations appearing in the CMO ensemble with the first-order filters. Figure 5.10 illustrates the changes in dynamic modes of the ensemble PPLL with the first-order filters at coupling parameter variations. In Fig. 5.10(a), full lines limit the regions of existence of the quasisynchronous mode I_{k2}. Inside these regions, the areas of chaotic oscillations of the type $[1, 0]$, that is chaotic oscillations corresponding to the generation of CMO at the outlet of the second ensemble generator, are marked dark. The boundaries of the regions H_{k2} are bifurcation curves of the fourth doubling of the stable limit cycle $L_{1,0}$. The curves corresponding to the first doubling of the cycle $L_{1,0}$ are dotted. As the system (5.3) with these parameters is invariant in relation to the substitute $(\kappa, \varphi_1, \varphi_2) \rightarrow (\delta, \varphi_2, \varphi_1)$, we can obtain the picture of the quasisynchronous mode regions I_{k1}, changing the axes designations $\kappa \leftrightarrow \delta$ in Fig. 5.10(a).

(a)

(b)

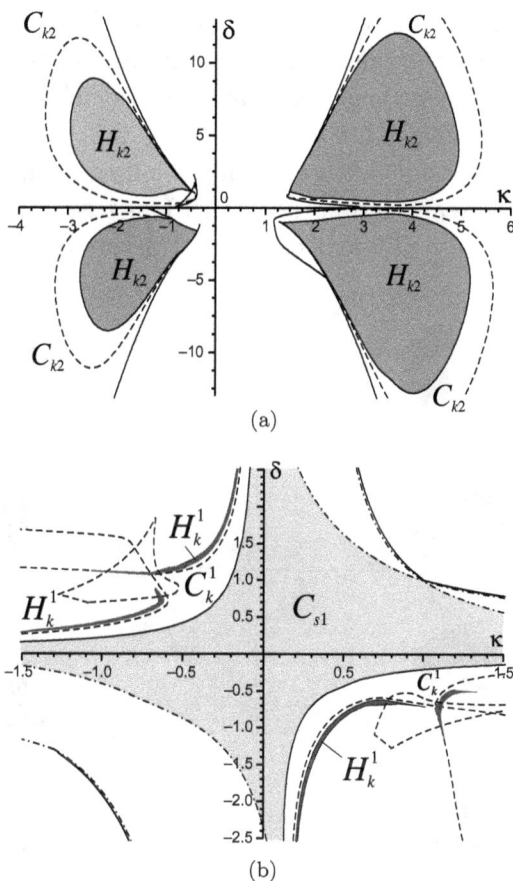

Fig. 5.10   Dependence of the dynamic processes of the model (5.3) on the couplings parameters at $\varepsilon_1 = \varepsilon_2 = 5$, $\gamma_1 = \gamma_2 = 0.3$.

Figure 5.10(b) shows the distribution of synchronous and quasisynchronous modes, to which stable equilibrium states and oscillatory attractors respond in the phase space of the model (5.3). In Fig. 5.10(b), the boundaries of the C_0 are marked with dash-and-dot lines, the Andronov–Hopf curve bifurcations are marked with full lines, the first bifurcations of doubling stable oscillatory cycle periods are marked with dotted lines, and the existence of the oscillatory-type chaotic attractors is marked dark. It should be noted that there are other chaotic oscillatory attractors existing on the plane κ, δ, but these are small, that is why in Fig. 5.10(b) they are not marked. So, if we combine the diagrams shown in Fig. 5.10, then a rather complicated figure of synchronous and quasisynchronous mode distribution will be obtained. Using this figure, it is possible to manage the ensemble oscillation features with the help of couplings. In order to illustrate the great possibilities of PPLL as a generator of CMO in Fig. 5.11 we can use different chaotic attractors of the model (5.3).

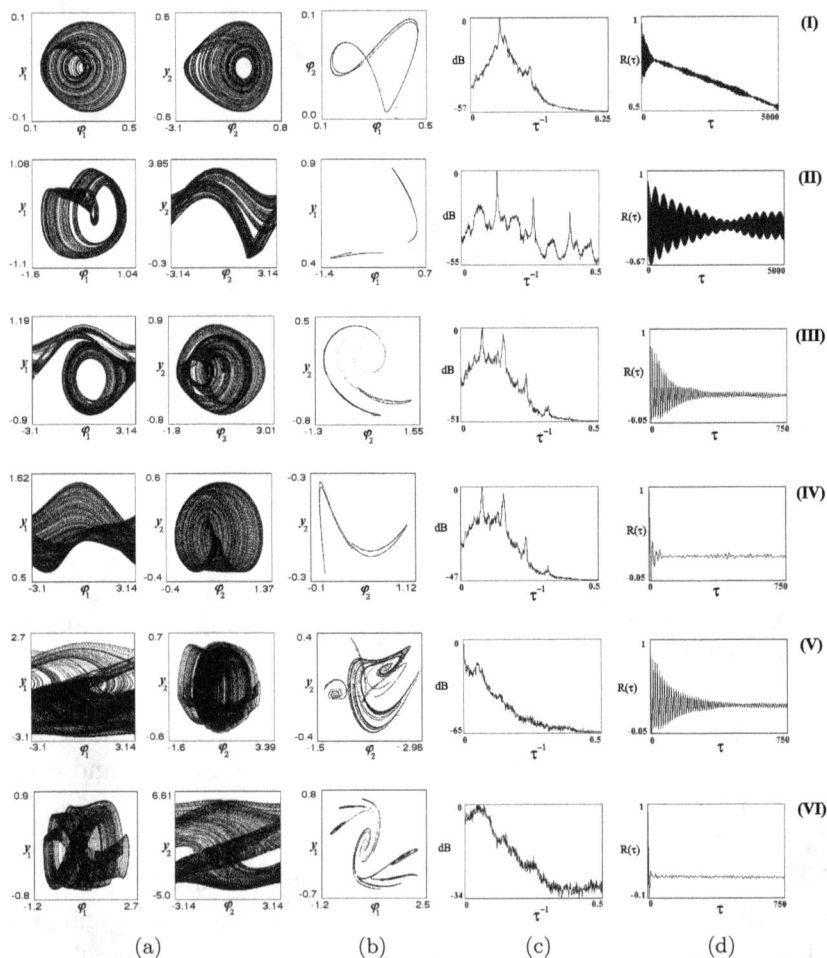

Fig. 5.11   The projections of (a) the phase portraits, (b) Poincaré mappings, (c) spectra and (d) autocorrelation functions of the attractor of the model (5.3).

The figure is created for the following parameters of the model (5.3):
(I) $\gamma_1 = 0.34$, $\gamma_2 = 0.33$, $\varepsilon_1 = 7.5$, $\varepsilon_2 = 5$, $\kappa = -0.1043$, $\delta = 3.07$;
(II) $\gamma_1 = 0.4$, $\gamma_2 = 1.8$, $\varepsilon_1 = 2.45$, $\varepsilon_2 = 0.6$, $\kappa = 1.73$, $\delta = 1.4$;
(III) $\gamma_1 = 0.36$, $\gamma_2 = 0.3$, $\varepsilon_1 = \varepsilon_2 = 5.0$, $\kappa = -0.7$, $\delta = 0.65$; (IV)
$\gamma_1 = 0.16$, $\gamma_2 = 0.3$, $\varepsilon_1 = \varepsilon_2 = 5.0$, $\kappa = -2.0$, $\delta = 0.73$; (V) $\gamma_1 = 0.7$,
$\gamma_2 = 0.3$, $\varepsilon_1 = \varepsilon_2 = 5.0$ $\kappa = 4.0$, $\delta = 0.7$; (VI) $\gamma_1 = 0.4$, $\gamma_2 = 1.8$,
$\varepsilon_1 = 2.45$, $\varepsilon_2 = 0.6$, $\kappa = 1.5$, $\delta = 5.4$. Item (I) corresponds to
the oscillatory attractor with the index $[0, 0]$; Items (II) and (VI)
correspond to rotatory–oscillatory attractors with the indices $[0, 1]$;
Items (III)–(V) correspond to rotatory–oscillatory attractors with
the indices $[1, 0]$.

5.3.2. *The second-order filter case*

Let us examine the evolution of dynamic modes of the ensemble PLL
with the second-order filters in the case when the coupling parameters change. While there are no couplings, let PLL_1 have the parameters $\gamma_1 = 0$, $\varepsilon_1 = \mu_1 = 1$, which belong to the region of pull-in to the
synchronous mode, and in PLL_2, the parameters $\gamma_2 = 0.69$, $\varepsilon_2 = 1$,
$\mu_2 = 2.37$ can function in one of the two regular modes: in the mode
of beatings, which corresponds to the rotatory limit cycle L_0 in the
phase space or in the quasisynchronous mode, which corresponds to
an oscillatory cycle $L_{01}^{(2)}$. We will combine these systems with the
help of the unidirectional coupling δ. At such coupling, the first PLL
dynamics is not changed, but the second PLL behavior coincides with
the dynamics of the partial PLL — the initial detuning of which is
$\gamma_2(\delta) = \gamma_2 - \gamma_1\delta$. As in the case under consideration $\gamma_1 = 0$, the
dynamics of PLL_2 at variations δ does not change. Further, we will
examine the evolution of dynamic modes of the ensemble generators
when coupling κ is introduced at different levels of δ.

Let us assume that $\delta = 0.9$. The possible scenarios of the
development of ensemble dynamic modes are shown in Fig. 5.12.
Figure 5.12(a) demonstrates the situation when there is a quasisynchronous mode in the second system. As κ grows, the PPLL dynamics develops in the following way. At $\kappa = 10^{-5}$ in the phase space
of the model (5.2), there is an oscillatory invariant torus T_0, which

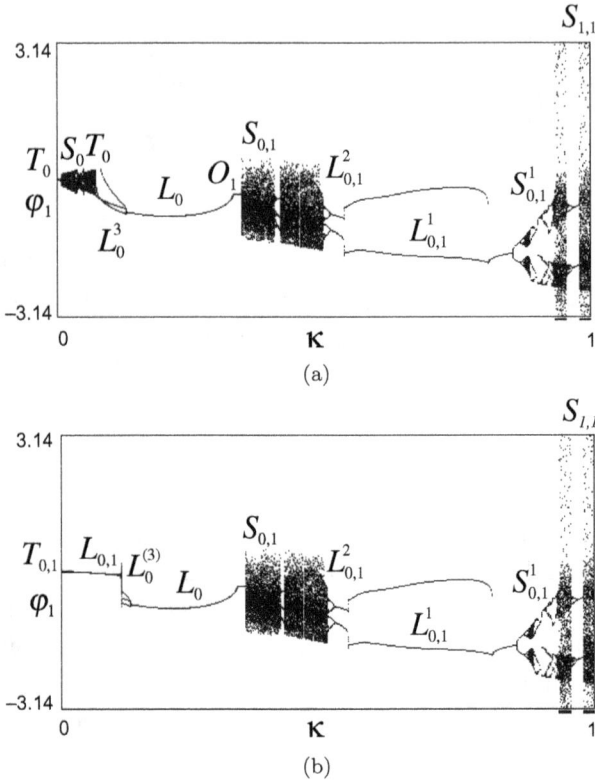

$$S_{1,1}$$

3.14

T_0 $S_0 T_0$ $S_{0,1}$ $L^2_{0,1}$

φ_1 L_0 O_1 $S^1_{0,1}$

L^3_0 $L^1_{0,1}$

−3.14

0 κ 1

(a)

$$S_{1,1}$$

3.14

$T_{0,1}$ $L_{0,1}$ $L^{(3)}_0$ $S_{0,1}$ $L^2_{0,1}$

φ_1 L_0 $S^1_{0,1}$

$L^1_{0,1}$

−3.14

0 κ 1

(b)

Fig. 5.12 Figures of developing of dynamic modes of the model (5.2) in case $\gamma_1 = 0$, $\varepsilon_1 = \mu_1 = 1$, $\gamma_2 = 0.69$, $\varepsilon_2 = 1$, $\mu_2 = 2.37$, $\delta = 0.9$ with increasing κ.

corresponds to two-frequency global quasisynchronous mode. The generation of the two-frequency mode can be explained by the fact that the frequency $\omega_1 = 0.1$ of an oscillatory limit cycle in the second system is smaller than the frequency $\omega_2 = 1$ of an oscillatory exhaustive process in the first generator by an order of magnitude. An increase in κ leads to loss of smoothness of the invariant torus and the appearance of a chaotic oscillatory attractor S_0. If κ grows, the chaotic attractor is again transformed into the torus T_0, on which a triple cycle $L^{(3)}_0$ appears. When κ further increases, the cycle $L^{(3)}_0$ disappears through the saddle-node bifurcation, and the model (5.2) moves to the oscillatory cycle L_0, which deforms to point O_1 with growing κ. The equilibrium state O_1 disappears at $\kappa = 0.328$, as a result, the system (5.2) moves to the chaotic attractor $S_{0,1}$ of the type [0, 1]. If κ increases, the attractor $S_{0,1}$ becomes regular, then becomes again chaotic.

At $\kappa = 0.98$, there is a crisis of the attractor $S_{0,1}$, and it transforms into a chaotic attractor of the rotatory type. The diagrams in Fig. 5.12 are created to show the changing κ on the interval from 0 to 1 in the case of "weak couplings". However, the examined interval of the changing κ depending on the coupling force and the type of a dynamic mode can be divided into several intervals: *very weak "weak coupling"* — regular beatings of the frequencies of modulated oscillations between the frequencies ω_1 and ω_2; *weak "weak coupling"* — chaotic beatings between ω_1 and ω_2; *medium "weak coupling"* — synchronization of modulating oscillations of the first and the second PLL on the definite rational correlations ω_1/ω_2; *strong "weak coupling"* — disappearance of modulating oscillations; *stronger "weak coupling"* — break to the mode of CMO, stabilized by the reference signal in the first generator and to chaotic beatings in the second generator; *very strong "weak coupling"* — chaotic beatings in both generators.

Figure 5.12(b) illustrates the evolution of dynamic modes if the mode of beating is initially realized in the second PLL. Here, if a very weak coupling $\kappa = 10^{-7}$ is introduced in the phase space, a stable invariant torus $T_{0,1}$ appears, which encircles the phase cylinder U along the coordinate φ_2. Now, two-frequency quasisynchronous mode is realized only in the first ensemble generator, the second generator functions in two-frequency mode of beatings. At $\kappa = 2 \times 10^{-6}$, the torus $T_{0,1}$ moves to the stable limit cycle $L_{0,1}$, which is kept till $\kappa = 0.1088$. At $\kappa = 0.1088$, the cycle $L_{0,1}$ disappears from a saddle-node bifurcation, and the system (5.2) moves onto the oscillatory cycle $L_0^{(3)}$. PPLL dynamics develops further to the above-described scenario.

Figure 5.13(a) shows the development of dynamic processes when κ grows, and when $\delta = 0.5$ in case a quasisynchronous mode is initially realized in the second system. Here, in contrast to the diagram in Fig. 5.12(a), at $\kappa = 10^{-5}$ there is a stable oscillatory limit cycle L_0. The cycle L_0 loses stability, when the increasing κ creates

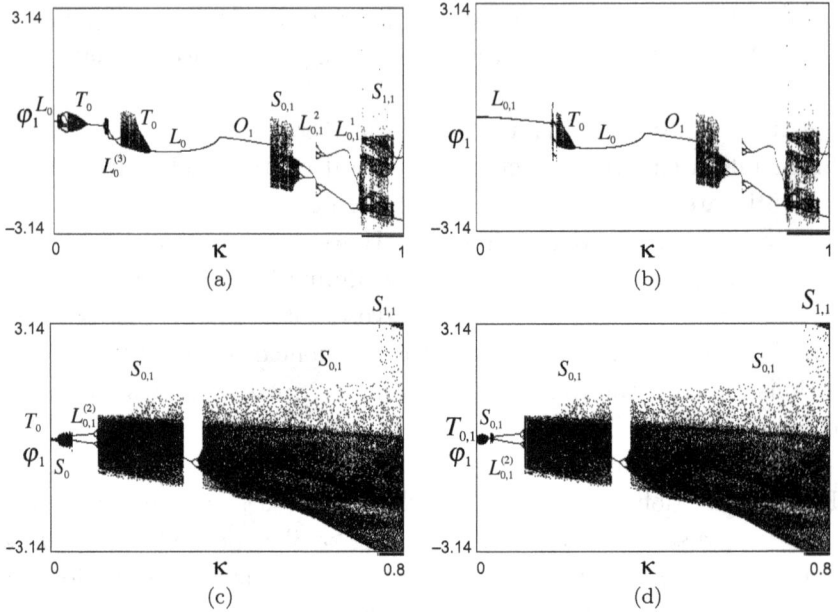

Fig. 5.13 Pictures of development of dynamic modes of the model (5.2) in case $\gamma_1 = 0$, $\varepsilon_1 = \mu_1 = 1$, $\gamma_2 = 0.69$, $\varepsilon_2 = 1$, $\mu_2 = 2.37$, (a, b) $\delta = 0.5$ and (c, d) $\delta = -0.5$ at increasing κ.

the invariant torus T_0, which at first is transformed into the chaotic attractor S_0, then into an oscillatory cycle and then again becomes a torus. Then the torus T_0 is transformed into a limit cycle which is deformed into a point O_1. In this case, the conversion process is accompanied with resonance cycles appearing and disappearing on the torus. The equilibrium state O_1 disappears at $\kappa = 0.619$, and the system (5.2) moves to the rotatory–oscillatory chaotic attractor $S_{0,1}$. With increase in κ, the attractor $S_{0,1}$ at first becomes regular, then disappears, and the system (5.2) comes to the chaotic attractor $S_{0,1}$ of the rotatory type. Figure 5.13(b) illustrates the evolution of dynamic modes when the mode of beatings is initially realized in the second PLL. Here, the scenario of dynamic mode development is analogous to the scenario shown in Fig. 5.12(b).

Figures 5.13(c) and 5.13(d) show the diagrams for $\delta = -0.5$ in cases when the quasisynchronous mode (Fig. 5.13(c)) and the beating mode (Fig. 5.13(d)) are initially realized in the second system. This diagram analysis shows that qualitatively the development scenarios for the dynamic modes when $\delta = -0.5$ do not practically differ from the ones examined above when $\delta > 0$. However, there are some quantitative changes, for example, in Figs. 5.13(c) and 5.13(d), the regions of the oscillatory attractors become narrower and regions of rotatory–oscillatory attractors become wider, and the break to a global mode of beatings occurs for weaker coupling κ.

The diagrams in Figs. 5.12–5.13 show great possibilities of managing dynamic modes in the ensemble with second-order filters at the expense of the coupling parameters. Oscillations of different complexity can be obtained by varying the coupling parameters. When both systems have simple dynamics, it is possible to change the nature of self-oscillations and also to broaden the regions of oscillations of a certain type.

Thus, as it can be seen from the above given results, the time lag effect introduction into the local control circuits leads to the complication in the PPLL ensemble behavior. It is connected with the appearance of new complex motions and, first of all, chaotic ones, which results in the increase in the multistability level. However, it should be noted that nonlinear couplings play the key role in the generation of complex motions. It is proved by the following facts: stability of the synchronous mode of PPLL with weak couplings at any parameter values of the first-order filters remains stable or chaotic attractor features remain stable within broadbands of parametric variations ε_1 and ε_2.

5.4. Regions of Self-Modulated Modes Generation

This section analyzes the regions of self-modulated modes in the parametric space. The section focuses on the regions of generation of CMO and the adjacent regions. The research is based on construction of dynamic mode maps of models (5.3) and (5.2), one-parameter bifurcation diagrams of Poincaré mapping, and the calculation of different characteristics of modulating oscillations. The obtained results are compared with the analogous study of cascade-coupled models.

As the models under consideration are multistable, they have several different maps at the same fixed parameters. The type of these maps is determined by the development algorithm, which assumes the following: firstly, the initial system state definition, secondly, the way active model parameters vary, and finally, the system state algorithm at transition from one parameter space point to another. In order to analyze the generation regions of the PPLL ensemble self-modulated modes, we used the algorithms, which allowed us to estimate the regions of quasisynchronous modes existence and pull-in. The algorithms are analogous to those used in the study of the cascade-coupled PLL.

5.4.1. The first-order filter ensemble

Figure 5.14 shows possible distributions of dynamic modes of the model (5.3) on the plane (γ_1, γ_2). In creating the map in Fig. 5.14(a), the starting state of the model (5.3) had parameter values $\gamma_1 = \gamma_2 = 0$, and phase variables belonged to the oscillatory limit cycle L_0. The active parameters varied according to the further described algorithm. At first, the parameter γ_1 was changed from zero to -0.5 and from zero to 1.5 with the step $\Delta_{\gamma 1} = 0.01$. In this case, the latest values of phase variables from the preceding step by γ_1 were used for the

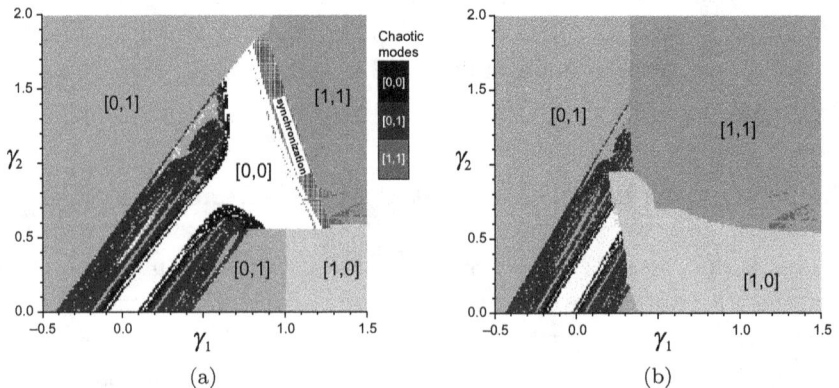

Fig. 5.14   Maps of dynamic modes of the model (5.3) if the couplings parameters $\varepsilon_1 = \varepsilon_2 = 5$, $\delta = 1.5$, $\kappa = -0.3$ characterizing the pull-in and pull-out regions of the synchronous and quasisynchronous modes.

succeeding value γ_1 as the initial state of system (5.3). Upon reaching the interval ends, the value γ_1 was placed back at the beginning of the intervals ($\gamma_1 = 0$) with restoration of the appropriate phase variables for this point, after that a step along $\gamma_2 = \gamma_2 + 0.01$ was made. Further, the procedure of varying γ_1 was repeated and stopped upon reaching $\gamma_2 = 2$. The map created as described above in Fig. 5.14(a) reflects the evolution of the mode I_{k1} at initial frequency detuning increase, i.e., it characterizes the pull-out region of the quasisynchronous mode I_{k1}. The diagram demonstrates that as γ_1 (γ_2) grows, a global regular quasisynchronous mode becomes chaotic. After that, it moves to the partial quasisynchronization mode of the type [0, 1], which later becomes regular. As in the case of the increased initial frequency detuning, the regular mode I_{k1} smoothly moves to the chaotic mode of global synchronization and further to the chaotic mode of partial synchronization; Fig. 5.14(a) characterizes the pull-out region of chaotic quasisynchronous modes of the model (5.3). Figure 5.14(a) shows that regions of chaotic quasisynchronous modes stretch along one of the diagonals of the parallelogram C_0, in case of greater initial detuning the oscillations are regularized. The regions in Fig. 5.14(a) occupy, respectively, 11.0%, [1, 1] — 0.3%, regular motions of the type [0, 0] — 14.5%, [0, 1] — 45.7%, [1, 0] — 7.1%, [1, 1] — 15.1%, the synchronization mode 3.5%. The parameter region, where the first ensemble generator functions in the generation mode of CMO, occupies totally 2.8% + 11.0% = 13.8%, but this region consists of two subregions which are divided by the region of regular oscillations. Thus, the index of chaotic homogeneity was calculated for each subregion separately: above (to the left) of the point ($\gamma_1 = 0, \gamma_2 = 0$) there is the index of chaotic homogeneity $I_1^1 = 0.8$, and for the subregion situated lower (to the right) it is $I_1^2 = 0.84$. The parameter region, where the second ensemble generator functions in the CMO generation mode, is not of great interest due to its small dimensions. The region of existence of the synchronous mode C_{S1} is located in the region of greater initial frequency detuning.

The map in Fig. 5.14(b) characterizes the pull-in regions of in quasisynchronous modes. Here, the point $\gamma_1 = 1.5$, $\gamma_2 = 2$ was used as the initial state, the phase variables belonged to the attractor of the rotatory type, the parameter γ_1 was changed first, γ_2 was changed next. In the course of movement along γ_1, the coordinates of phase variables from the previous step were used as the initial state of the system (5.3). In the course of movement along γ_2, the values γ_1 returned to the point $\gamma_1 = 2$ with the restoration of phase variable values corresponding to the point. The regions in Fig. 5.14(b) occupy, respectively, chaotic motions of the type $[0, 0]$ — 1.0%, $[0, 1]$ — 6.5%, $[1, 1]$ — 0.3%, regular motions of the type $[0, 0]$ — 2.5%, $[0, 1]$ — 30.0%, $[1, 0]$ — 20.0%, $[1, 1]$ — 39.7%, the mode of synchronization 0.3%. Comparing the regions in Figs. 5.14(a) and 5.14(b), we can see that there is no synchronous mode pull-in region at the parameter values under consideration, and the CMO mode pull-in regions are half the size of the pull-out regions.

The map in Fig. 5.15(a) depicts the structure of the CMO region at the outlet of the first generator in the parameter plane $(\varepsilon_1, \gamma_1)$. In order to create the map the parameter values $\gamma_1 = 0$, $\varepsilon_2 = 1$ were the starting state of the model (5.3). The phase variables belonged to the rotatory–oscillatory attractor of the type $[0, 1]$. The parameter γ_1 was the first active parameter. It was changed from zero to ± 0.5 with the step $\Delta_{\gamma 1} = 0.01$. The phase variable values for the next parameter γ_1 were inherited from the attractor in the previous step. The parameter ε_1 varied in the logarithmic scale from 0 to 3 with the step $\Delta_{\lg(\varepsilon 1)} = 0.01$. In this map, dynamic modes are distributed in the following way: chaotic motions of the type $[0, 1]$ occupy 52%, $[1, 1]$ — 4%, regular motions of the type $[0, 0]$ occupy 2%, $[0, 1]$ — 40%, $[1, 1]$ — 2%. The index of chaotic homogeneity of the CMO generation region at the outlet of the first generator is equal to $I_1 = 0.94$. Figure 5.15(b) represents the map of the maximum Lyapunov exponent. It is clear that the chaotic degree of the attractor $S_{0,1}$ with increasing ε_1 becomes lower. The comparison of chaotic motion of the $[0, 1]$ PPLL-type and CPLL shows that in case of parallel coupling the chaotic degree of these motions is higher, the region of existence of the chaotic attractor $S_{0,1}$ along ε_1 is more stretched, the index of chaotic homogeneity of this region is higher.

Figure 5.16 illustrates the features of the first PLL modulating oscillations in case of changing parameters ε_1 and γ_1. It is clear that

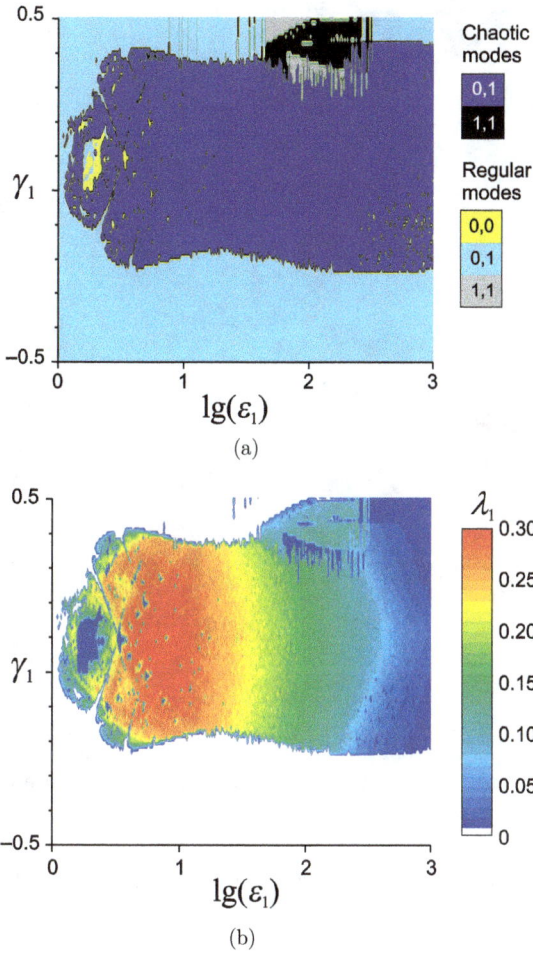

Fig. 5.15   Maps of (a) dynamic modes (5.3) and of (b) the maximum Lyapunov exponent of the model (5.3) for $\gamma_2 = 0.3$, $\varepsilon_2 = 5$, $\delta = 3$, $\kappa = -1$.

the averaged values $\tilde{y}_1$ practically do not depend on the parameters of PLL$_1$, the values of $\tilde{\varphi}_1$ increase with the increase in γ_1, if ε_1 changes they do not change. The ranges of oscillations $\Delta\varphi_1$ and Δy_1 decrease if ε_1 grows. If $\varepsilon_1 > 100$, oscillations y_1 weaken.

Figure 5.17 shows the distribution of dynamic modes on the plane of parameter couplings. The point $\delta = 5.2$, $\kappa = -0.1$ and oscillatory–rotatory limit cycle of the type $[0, 1]$ were the starting state for the

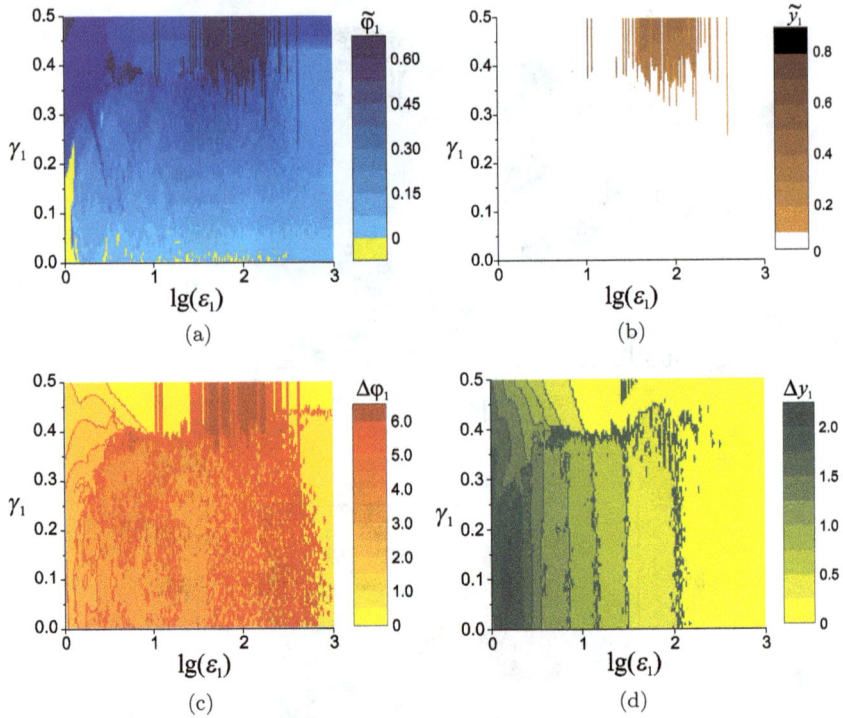

Fig. 5.16   Diagrams of evolution of these modulating oscillations characteristics PLL_1 of the ensemble of two PLLs (PPLL) connected in parallel for $\gamma_2 = 0.3$, $\varepsilon_2 = 5$, $\delta = 3$, $\kappa = -1$.

creation of the map. The parameter δ was chosen as the first active parameter; it was changed from 5.2 to 0 with the step $\Delta_\delta = 0.02$, and the second active parameter κ was changed from -0.1 to -5.2 with the step $\Delta_\kappa = 0.02$. The latest values of phase variables from the preceding step were used as the initial state of the model (5.3) in moving along the parameter. The principle components of the parametric portrait in Fig. 5.17 are regions of chaotic motions of the type $[0, 1]$ — 16%, $[1, 0]$ — 16%, $[1, 1]$ — 62%. The other motion regions occupy a small part here: motions of the type $[0, 0]$ occupy 2.4%, $[0, 1]$ — 1.4%, $[1, 0]$ — 2%, synchronous mode 0.2%. The regions of CMO generation at the outlet of the first ensemble generator stretch along the X-axis and the regions of CMO generation at the outlet of the second ensemble generator stretch along the Y-axis. The transition

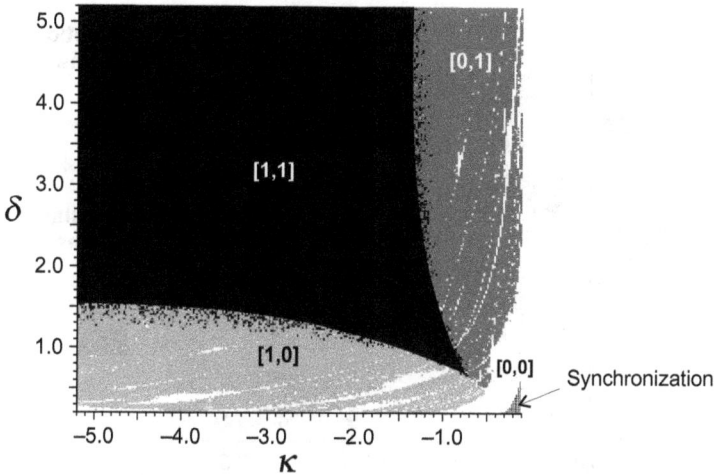

Fig. 5.17 The maps of dynamic modes of the model (5.3) for $\gamma_1 = \gamma_2 = 0.1$, $\varepsilon_1 = \varepsilon_2 = 10$.

from one region of CMO generation to another is performed through the chaotic beating region. It is hard to draw a clear, cut boundary line between the CMO generation and chaotic beating regions, as the generation of chaotic beating is connected with thin turns starting on the oscillatory–rotatory attractors on 2π of the variable φ_1 in the attractors of the type [0, 1] and φ_2 in the attractors of the type [1, 0]. Moving further into the region of chaotic beatings the turns along φ_1 and φ_2 become equiprobable. The CMO generation in Fig. 5.17 has rather high indices of chaotic homogeneity: at the outlet of the first generator $I_1 = 0.92$, and at the outlet of the second generator $I_2 = 0.87$.

5.4.2. *Ensemble with the second-order filters*

Figure 5.18(a) shows the map of dynamic modes of the model (5.2) on the parameter planes (μ_1, γ_1), created according to the algorithm, which allows us to estimate the pull-out regions of a quasisynchronous mode. As the starting state for the creation of this map, we chose the values of the parameters $\gamma_1 = 0$, $\mu_1 = 0.01$ and the phase variables on the attractor of the type [0, 0]. Further, the initial

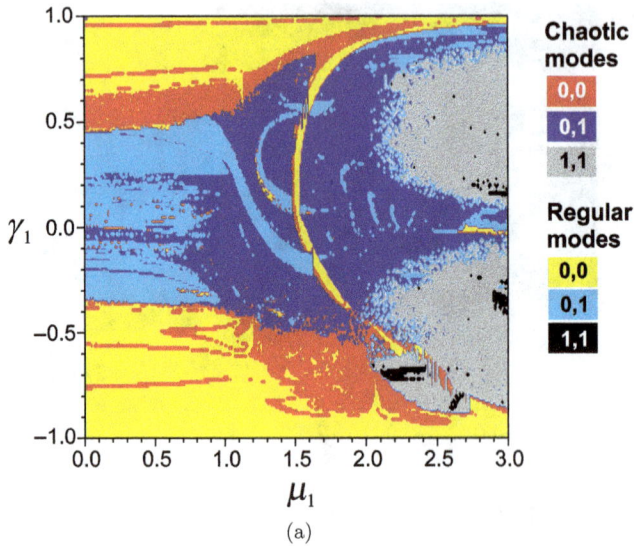

(a)

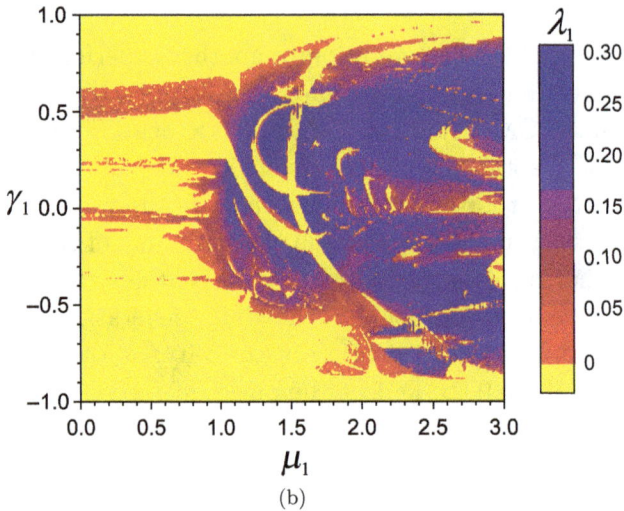

(b)

Fig. 5.18 Maps of (a) dynamic modes and of (b) the maximum Lyapunov exponent of the model (5.2) for $\gamma_2 = 0.39$, $\varepsilon_1 = \varepsilon_2 = 1$, $\delta = -0.6$, $\kappa = 0.1$.

frequency detuning was increased with the step $\Delta_{\gamma 1} = 0.01$ until a global mode of beatings was installed in the system. In this case, the phase variable values for the next parameter value γ_1 were inherited from the attractor in the previous step. Then the described procedure was repeated for values $\mu_1 \in [0.01, 3.0]$ with the step $\Delta_\mu = 0.01$. The obtained map consists of the following regions: chaotic motions of the type $[0, 0]$ occupy 8%, $[0, 1]$ — 28%, $[1, 1]$ — 15%, regular motions of the type $[0, 0]$ occupy — 32%, $[0, 1]$ — 15%. It is clear that the regions of CMO generation, as well as at the cascade coupling of the similar systems, are extremely heterogeneous. The chaotic regions alternate with regular motion regions. The chaotic homogeneity index of the region of CMO generation at the outlet of the first generator is $I_1 = 0.71$. The region of CMO generation by the second generator of the type $[0, 0]$ consists of two subregions. One subregion is situated in the region of positive γ_1 and occupies 4.5% of the map area, its chaotic homogeneity index is $I_2^1 = 0.96$, the other is situated in the region of negative γ_1, it occupies 5.9% and has the index $I_2^2 = 0.88$. Figure 5.18(b) shows the map of the maximum Lyapunov exponent. The combined analysis of Figs. 5.18(a) and 5.18(b) testifies that the oscillatory–rotatory type attractors have the biggest Lyapunov exponent. In general, chaotic motion region characteristics of the represented map are better than the partial PLL with the second-order filter. For PPLL, they can be slightly improved at the expense of the coupling parameter δ; however, according to the research, these improvements are not considerable. The maps of the dynamic ensemble modes of the systems with the second-order filter are extremely heterogeneous.

Let us examine the ensemble dynamics of two parallel coupled PLL, when the individual dynamics of PLL_1 can only be regular and PLL_2 has chaotic dynamics. For this purpose, the first-order filter is chosen in the first generator's control circuit, and the second-order filter is chosen in the second generator's control circuit. Let the parameters PLL_1 belong to the pull-in region of a synchronous mode, and the individual dynamics of PLL_2 is characterized by the mode of chaotic beatings. We will introduce the coupling $\delta = 0.9$. Such coupling degree leaves PLL_2 in the mode of chaotic beatings, and the unidirectional coupling δ has no influence on the dynamics of PLL_1. The dynamics of PLL_1 will change if the coupling κ is switched on; in this case if $\delta \neq 0$, the variations κ lead to the change in the dynamics of both ensemble generators.

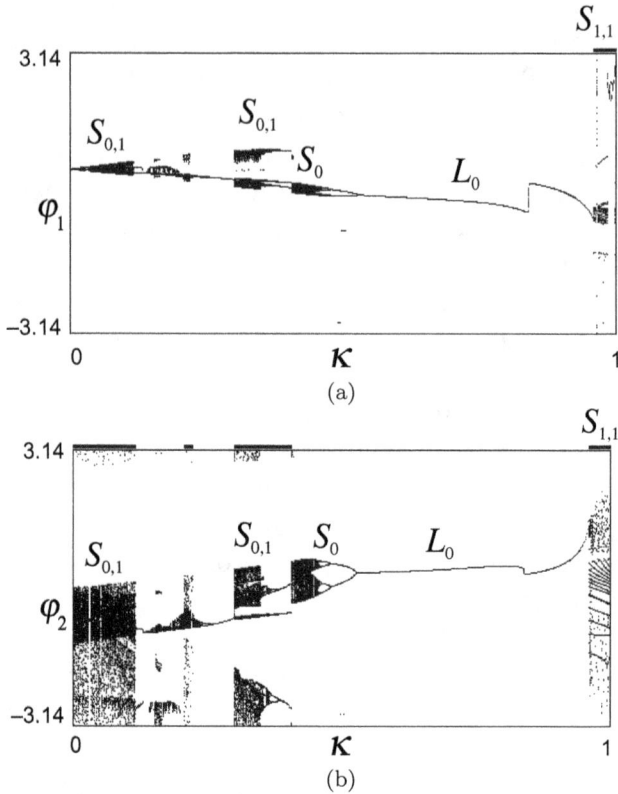

Fig. 5.19 The development of the model dynamic modes (5.2) at $\gamma_1 = 0.5$, $\varepsilon_1 = 1, \mu_1 = 0, \gamma_2 = 0.5, \varepsilon_2 = 1, \mu_2 = 6, \delta = 0.9$ if κ increases.

Figure 5.19 illustrates the changes in the ensemble generator dynamics caused by the introduction of the positive coupling κ if $\delta = 0.9$. In the interval $K_0 : (0; 0.118)$ in the phase space of the model (5.2), the chaotic oscillatory–rotatory attractor $S_{0,1}$ is realized corresponding to CMO generation at the outlet of the first generator and chaotic beatings of PLL_2. Then there are the following intervals: $K_1 : (0.118; 0.206)$ — the interval of global quasisynchronization, $K_2 : (0.206; 0.226)$ — the interval of partial chaotic quasisynchronization of the type $[0, 1]$, $K_3 : (0.226; 0.302)$ — the interval of global quasisynchronization, $K_4 : (0.302; 0.408)$ — the interval of partial quasisynchronization of the type $[0, 1]$,

$K_5 : (0.408; 0.462)$ — the interval of global chaotic quasisynchronization, $K_6 : (0.462; 0.838)$ — the interval of global regular quasisynchronization, $K_7 : (0.838; 0.956)$ — the interval of global synchronization. If $\kappa > 0.956$ in the ensemble, a global mode of chaotic beatings starts. In the intervals K_1, K_3, K_4, motions can be either regular or chaotic. Thus, the represented results confirm the previously drawn conclusion concerning the possibility of managing the ensemble oscillation features at the expense of insignificant changes in coupling parameters.

Now, we will analyze the ensemble dynamic modes on the parameter planes $\{\varepsilon_1, \gamma_1\}$ in cases when $\kappa \in K_0$ and $\kappa \in K_5$ (see Fig. 5.20). The maps in Figs. 5.20(a) and 5.20(c) are obtained when the initial frequency detuning γ_1 increases from the zero values, when

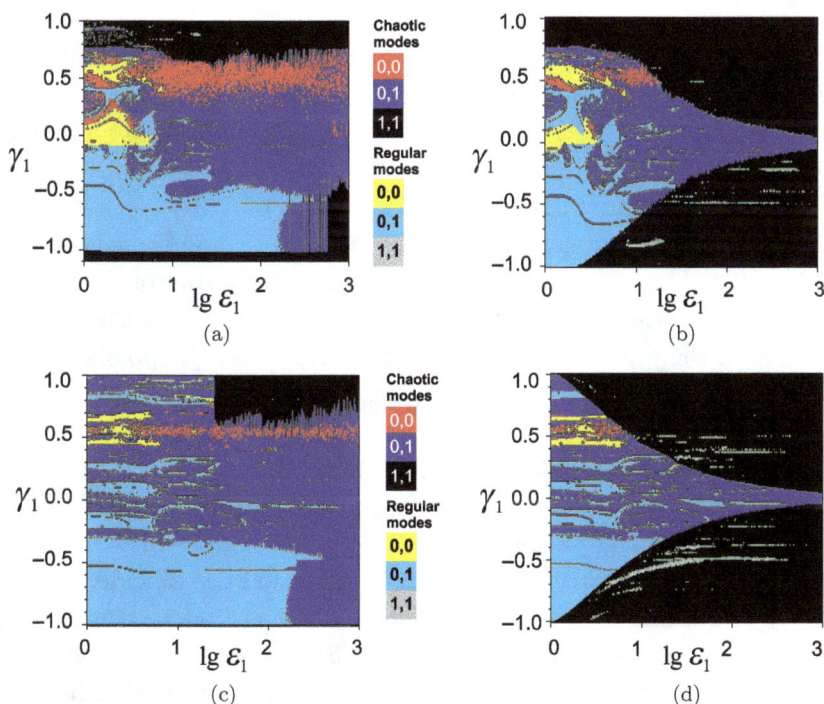

Fig. 5.20 Maps of dynamic modes (5.2) for $\mu_1 = 0$, $\gamma_2 = 0.5$, $\varepsilon_2 = 1$, $\mu_2 = 6$, $\delta = 0.9$, (a, b) $\kappa = 0.47$ and (c, d) $\kappa = 0.1$.

a quasisynchronous mode is realized in the ensemble, to the values $\gamma_1 = \pm 1$, when a global mode of beatings is realized in the ensemble. In general, these maps reflect the process of pull-out of a quasisynchronous mode. The diagrams in Figs. 5.20(b) and 5.20(d) are created for γ_1 decreasing from $\gamma_1 = \pm 1$ to 0 and reflect the process of pull-in to a quasisynchronous mode.

The analysis of the given diagrams allows us to make the following conclusions. In Fig. 5.20(a), dynamic modes are distributed in the following way: chaotic motions of the type [0, 0] occupy 10%, [0, 1] — 43%, [1, 0] — 4%, [1, 1] — 11%, regular motions of the type [0, 0] and [0, 1] occupy 5% and 27%, respectively. In Fig. 5.20(b), chaotic motions of the type [0, 0] occupy 1.3%, [0, 1] — 21.2%. The regions of CMO generation at the outlet of the first generator occupy 52% (its chaotic homogeneity index is $I_1 = 0.87$), at the outlet of the second generator, they occupy 7% (its chaotic homogeneity index is $I_2 = 0.82$).

In Fig. 5.19(c), the regions of dynamic modes are distributed in the following way: chaotic motions of the type [0, 0] occupy 2%, [0, 1] — 49% [1, 0] — 3% [1, 1] — 15%, regular ones of the type [0, 0] occupy 3%, [0, 1] — 28%. In Fig. 5.20(d), chaotic motions of the type [0, 0] occupy 0.5%, [0, 1] — 19%. The region of CMO generation at the outlet of the first generator is 51%, its chaotic homogeneity index is $I_1 = 0.8$. The region of CMO generation at the outlet of the second generator consists of the points of the parameter space, the chaotic motion of the type [0, 0] is set for them. However, the points of this type in Fig. 5.19(c) are extremely scattered and do not create any significant regions.

Thus, the results obtained in the course of our research testify to the following: firstly, the dynamic features of PPLL to a great degree depend on coupling parameters; secondly, the regions of CMO pull-out and the pull-in regions to CMO modes do not coincide, particularly it becomes obvious if the filter time constant is bigger, and finally, a decrease in the first generator filter order improves the chaotic homogeneity index of the CMO generation regions, but these regions still remain extremely heterogeneous, especially in the region of small filter's time constant.

In conclusion, the comparative analysis of the obtained results and the analogous research of CMO of the ensemble with a cascade type of coupling allow us to draw the inference that a cascade coupling for CMO generation is more preferable.

Chapter 6

Synchronization of Chaotic Oscillations

The phenomenon of chaotic synchronization was discovered about three decades ago [14, 16, 68] and stimulated numerous publications. However, a satisfactory theory of synchronization of chaotic oscillations has not been constructed until now. Moreover, interpretation of the obtained results in terms of their correlation with the classical synchronization of periodic oscillations revealed a great difference between the synchronization of chaotic oscillations and the synchronization of periodic oscillations. Some researchers believe that the term "synchronization" is not quite accurate for chaotic oscillations, a more relevant term would be, for instance, "coordination of chaotic oscillations", or "symmetry of chaotic oscillations", or "consistency of chaotic oscillations". Note also that the terms "chaotic synchronization" and "synchronization of chaotic oscillations" are usually regarded to be equivalent.

Different approaches to the problem of synchronization of chaotic oscillations are available in the literature. These include complete synchronization [14, 16, 68], generalized synchronization [69, 70], lag synchronization [71, 72], frequency synchronization [73], phase synchronization [74], and others. Such a diversity of terms and approaches speaks about difficulties in constructing a sound theory of synchronization of chaotic oscillations, greatly due to the fuzzy concepts of phase and frequency of chaotic oscillations and the involved indefiniteness of synchronization criteria, etc. It is also worthy to note that the current development in the theory of chaotic synchronization is not application oriented, hence many problems of chaotic synchronization are formulated based on the model approximations rather than following application-based requirements.

At the beginning of the 1990s, there appeared a bright idea to use dynamic chaos for secure communication of information [68]. The broadband nature of chaotic signals and relative simplicity of producing generators of chaos opened up challenging prospects for creating new competitive communication systems using dynamic chaos. These systems promised to ensure high-rate modulation, large information capacity, confidential transmission of information, and other opportunities. The idea proved to be very attractive. The theoretical researches and prototype developments [20, 23] confirmed that it is possible, in principle, to realize transmission of information using chaotic oscillations. However, in the receiver generating identical chaotic oscillations that could be used as carrier oscillations (which is of fundamental importance in coherent reception), i.e., attaining the effects of complete synchronization or synchronous response, the parameters of chaos generators in the transmitter and in the receiver are required to be highly identical. This results in weak noise stability and creep resistance in the developed communication systems [21]. The inevitable conclusion is that the creation of competitive coherent transmitter–receiver systems using dynamic chaos has no prospects [75]. An alternative approach to solve the problem is to use chaotically modulated oscillations (CMO) for communication of information [26]. The broadband features of chaotic signals are, generally, worse for CMO but the undoubted advantage is their reliable synchronization. In the previous chapters, we demonstrated that the features of collective dynamics of ensembles of coupled phase-locked loop (PLL) systems may be successfully employed for construction of high-efficiency generators of CMO with central frequency stabilized by the reference oscillation. In the next section, we will consider how such oscillations may be synchronized.

6.1.　Synchronization of CMOs from Phase Systems

Consider synchronization of chaotic oscillations of two generators of chaotic oscillations coupled unidirectionally [26, 59, 76]. We will first analyze PLL_1 as a master oscillator of CMO and PLL_2 as a synchronized oscillator of CMO. Parameters of PLL_1 and PLL_2 are chosen to be close but not the same. If PLL_1 and PLL_2 are not coupled, then under the action of the control circuits of the PLL systems, the central frequencies of CMO_1 and CMO_2 at the outputs of PLL_1 and PLL_2, respectively, are stabilized by the frequency of the reference signal. However, as PLL_1 and PLL_2 have different parameters (the systems are not identical), the CMO in PLL_1 differ from the CMO in PLL_2. Consequently, when PLL_1 and PLL_2 are coupled, for CMO_1 and CMO_2 to be synchronized, the generator of chaotic oscillations with angular modulation at the output of PLL_2 must follow the changes in chaotic oscillations with angular modulation at the output of PLL_1. This can be achieved within the framework of a coupled PLL_1 and PLL_2 system, provided that the speed of the control circuits in the PLL systems is higher than the characteristic times of modulation. Naturally, parameters of the control circuits should be chosen to ensure admissibly small following error.

As coupled PLL_1 and PLL_2 models are written for instantaneous phase and frequency differences of oscillations at the outputs of PLL_1 and PLL_2 relative to the reference signal and, in addition, as CMO_1 and CMO_2 at the outputs of PLL_1 and PLL_2 are quasiperiodic with chaotic angular modulation, CMO_1 and CMO_2 will be synchronized if the coordinates $x_1(t) = x_2(t)$ of the dynamic states of the two coupled PLLs coincide. By virtue of PLL_1 and PLL_2 non-identity, closeness of $x_1(t)$ and $x_2(t)$, i.e., fulfillment of $|x_1(t) - x_2(t)| < \varepsilon$ (ε is synchronization accuracy), may be the condition of synchronization.

For achieving synchronization, we introduce unidirectional couplings between PLL_1 and PLL_2 along the φ, y, z coordinates describing instantaneous states of the systems. A variant of unidirectional coupling of PLL_1 and PLL_2 through additional discriminator D_{12} is shown in Fig. 6.1. If D_{12} is a phase discriminator (PD_{12}), then the signal $\delta_\varphi \sin(\varphi_2 - \varphi_1)$ formed at the output of PD_{12} is used to control the frequency of oscillator G_2. If PLL_1 and PLL_2 are coupled through the additional frequency discriminator FD_{12}, then the signal $\delta_y \Phi(y_2 - y_1)$ formed at its output is used to control the frequency of oscillator G_2. The nonlinearity of FD may, generally,

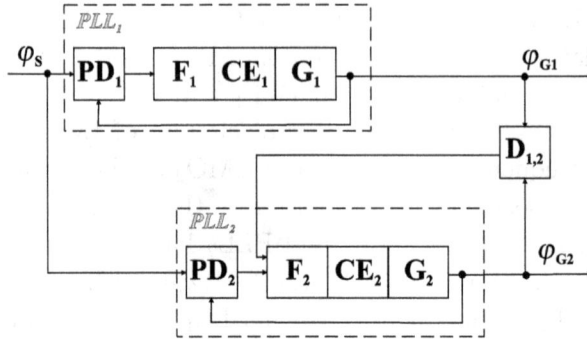

Fig. 6.1 Block diagram of synchronization of chaotic oscillations of G1 and G2.

be approximated by the function $\Phi(y) = 2ay/(1 + a^2y^2)$. To the first approximation, we can regard $\Phi(y)$ to be a linear characteristic and assume $\delta_y \Phi(y_2 - y_1) = \delta_y(y_2 - y_1)$. Finally, unidirectional coupling may also be realized along the z coordinate, so that the signal $\delta_z(z_2 - z_1)$ should arrive at the control circuit of G2. Introduction of such couplings (Fig. 6.1) is quite justified as controlling the PLL_2 oscillator by means of these couplings is aimed at tuning chaotic oscillations of PLL_2 oscillator to chaotic oscillations of PLL_1 oscillator.

Generally, for three control circuits along $\varphi, y,$ and z, the mathematical model of PLL_1 and PLL_2 coupled in this way is written in the form

$$\frac{d\varphi_1}{d\tau} = y_1, \quad \frac{dy_1}{d\tau} = z_1,$$

$$\mu_1 \frac{dz_1}{d\tau} = \gamma_1 - \sin \varphi_1 - y_1 - \varepsilon_1 z_1, \tag{6.1}$$

$$\frac{d\varphi_2}{d\tau} = y_2, \quad \frac{dy_2}{d\tau} = z_2,$$

$$\mu_2 \frac{dz_2}{d\tau} = \gamma_2 - \sin \varphi_2 - y_2 - \varepsilon_2 z_2$$

$$-\delta_z(z_2 - z_1) - \delta_y \Phi(y_2 - y_1) - \delta_\varphi \sin(\varphi_2 - \varphi_1).$$

For modeling dynamics of (6.1), the parameters $\gamma_{1,2}, \varepsilon_{1,2}, \mu_{1,2}$ of PLL_1 and PLL_2 are taken to be non-identical but rather close (the

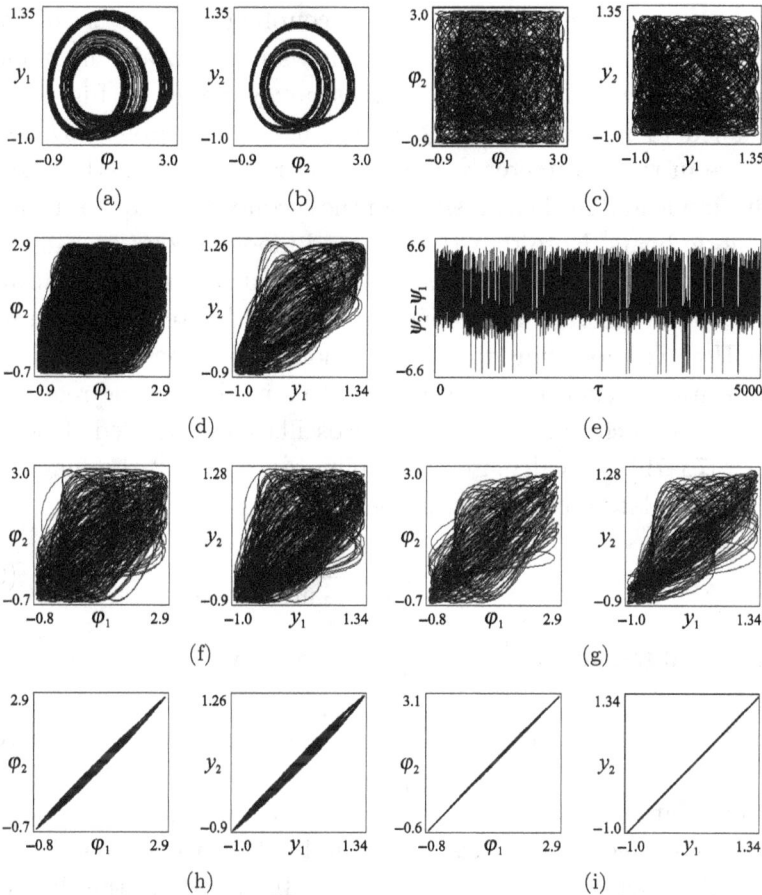

Fig. 6.2 Projections of chaotic attractor of the model (6.1)(a–c) for $\gamma_1 = 0.46$, $\gamma_2 = 0.51$, $\varepsilon_1 = \varepsilon_2 = 1$, $\mu_1 = \mu_2 = 2.2$, for $\delta_\varphi = \delta_y = \delta_z = 0$; (d) $\delta_\varphi = -0.02, \delta_y = \delta_z = 0$; for linear control (f) $\delta_\varphi = \delta_z = 0, \delta_y = -0.02$; (g) $\delta_\varphi = \delta_y = 0, \delta_z = 0.2$; (h) $\delta_\varphi = 0, \delta_y = 0.13, \delta_z = 0.2$; for nonlinear control (i) $\delta_y = 0.3, a = 30$ and the difference of phases ψ_1 and ψ_2 for (e) $\delta_\varphi = -0.02, \delta_y = \delta_z = 0$.

difference being 1–5%), and the initial states of PLL$_1$ and PLL$_2$ meet the following conditions: $y_1^0 = \gamma_1, y_2^0 = \gamma_2, z_1^0 = z_2^0 = 0$, φ_1^0, φ_2^0 are random quantities in the interval from $-\pi$ to π. The results of modeling dynamics of (6.1) are illustrated in Fig. 6.2.

In the absence of coupling $\delta_\varphi = \delta_y = \delta_z = 0$, chaotic oscillations of PLL$_1$ (Fig. 6.2(a)) and PLL$_2$ (Fig. 6.2(b)) are not synchronized

(Fig. 6.2(c)). In the model (6.1), weak coupling with the use of phase discriminator PD_{12} ($0 < |\delta_\varphi| < 0.1$) allows us to obtain the mode of weak convergence of phase coordinates φ_1, φ_2 and y_1, y_2 (Fig. 6.2(d)). We failed to improve convergence by means of coupling δ_φ, as an increase in the parameter δ_φ ($|\delta_\varphi| > 0.1$) resulted in the breakdown of this mode and in the transition of the second oscillator to the mode of chaotic beats. Note that the observed mode of convergence of the coordinates φ_1, φ_2 and y_1, y_2 cannot be regarded as a synchronization mode as the error of coincidence of the coordinates y_1, y_2 is rather high. However, following the classification of the types of chaotic synchronization known in the literature, this mode may be referred as the "phase synchronization of chaotic oscillations". Indeed, if, according to [76–79], we calculate at coupling $\delta_\varphi = -0.02$ the magnitude of the modulating oscillation phase

$$\psi_{1,2} = \arctan \frac{y_{1,2}}{\varphi_{1,2} - \arcsin \gamma_{1,2}} \qquad (6.2)$$

and central frequency

$$\Omega_{1,2} = \langle \dot{\psi}_{1,2} \rangle = \lim_{T \to \infty} = \frac{\psi_{1,2}(T) - \psi_{1,2}(0)}{T}, \qquad (6.3)$$

for modulating chaotic oscillations of $G_{1,2}$ the images of which are attractor projections in Figs. 6.2(a) and 6.2(b), the central frequencies will coincide: $\Omega_1 = \Omega_2 = -0.6429$. In this case, the difference between ψ_1 and ψ_2 (Fig. 6.2(e)) is essentially non-zero, though limited (phase pull-in).

Like in the previous case, the introduction of linear couplings $\delta_y \neq 0$ along $y = y_2 - y_1$ does not ensure a satisfactory synchronization mode as the error of coincidence of the difference of frequencies y_1, y_2 is still large enough, although the phase pull-in becomes more pronounced (Fig. 6.2(f)), i.e., only the central modulation frequencies are synchronized (chaotic phase synchronization). An analogous result is obtained when linear coupling $\delta_z \neq 0$ is introduced along $z = z_2 - z_1$ (Fig. 6.2(g)). Combined linear control of y and z (two linear couplings $\delta_y \neq 0$ and $\delta_z \neq 0$ are switched on simultaneously) may reduce the error of coincidence of the difference of phases φ_1, φ_2

and the difference of frequencies y_1, y_2 (Fig. 6.2(h)), although, e.g., the phase error is still quite pronounced $|\varphi_2 - \varphi_1| < \epsilon \simeq 0.4$.

Finally, the best result is attained by means of nonlinear control along the $y = y_2 - y_1$ coordinate, i.e., with nonlinear coupling of the form $\delta_y \Phi(y)$ (Fig. 6.2(i)). By varying the coupling force, it is possible to reduce synchronization error down to small values (for $a = 30$, $\delta_y = 0.3$, $\epsilon = 0.07$; $\delta_y = 0.6$, $\epsilon = 0.03$; $\delta_y = 1$, $\epsilon = 0.01$), i.e., it is possible to synchronize CMO_1 and CMO_2 to a needed accuracy at the output of PLL_1 and PLL_2, respectively.

Let us now analyze the obtained processes of chaotic synchronization. Apparently, it is reasonable to carry out the consideration in terms of using dynamic chaos for information transmission. In this case, the problems of chaotic synchronization in a general formulation almost do not differ from the problems of chaotic synchronization of regular oscillations. Like in the case of regular oscillations, they include assessment of the quality of synchronization, specifically, assessment of the size of the pull-in and pull-out regions of chaotic synchronization, synchronization accuracy, and the onset time of the synchronization mode.

The mode of chaotic synchronization. In the phase space of the model (6.1) a stable manifold Υ^S — a synchronization manifold — corresponds to the synchronization mode. If chaotic oscillations of identical systems are synchronized (in the considered case, this means identity of PLL_1 and PLL_2 parameters: $\gamma_1 = \gamma_2$, $\varepsilon_1 = \varepsilon_2$, and $\mu_1 = \mu_2$ in the model (6.1)), the projections of Υ^S on the plane of the corresponding phase variables (φ_1, φ_2), (y_1, y_2), and (z_1, z_2) are sections on the diagonals. The projections of Υ^S on the difference plane of the corresponding phase variables $[(\varphi_2 - \varphi_1), (y_2 - y_1)]$, $[(\varphi_2 - \varphi_1), (z_2 - z_1)]$, and $[(y_2 - y_1), (z_2 - z_1)]$ are points. For non-identical systems, the projections of the manifold Υ^S on the plane of the corresponding phase variables always have a "thickness" (Figs. 6.3(a–c)), and on the difference planes of the corresponding phase variables, certain manifold resembling projections of an oscillatory chaotic attractor appear (instead of a point) with phase variables changing within small limits (Figs. 6.3(d–f)). It is worthy to note that for non-identical systems, some projections

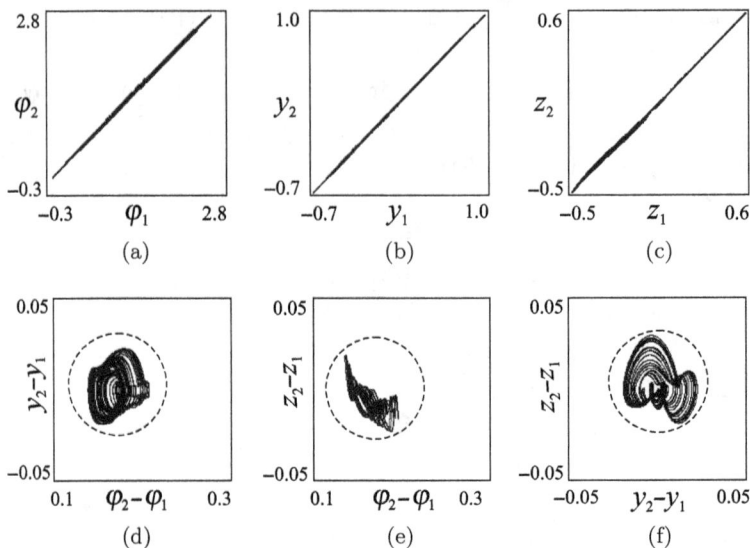

Fig. 6.3 Projections of the synchronization manifold Υ^S for identical systems. The dashed line reproduces ϵ circumference characterizing synchronization accuracy.

of Υ^S on the plane of the corresponding phase variables may lie not in the neighborhood of the diagonals but in the neighborhood of straight lines parallel to the diagonals as shown in Fig. 6.3(a). In this case, synchronization occurs to an accuracy of the constants that depend on the magnitude of the shift of the straight lines relative to the diagonals. The values of the constants determine the residual error of synchronization by the corresponding variables.

Chaotic synchronization accuracy. The thickness of the manifold Υ^S or the ϵ-radius circumference that may enclose the projection of Υ^S on the difference planes of phase variables characterizes synchronization accuracy. This characteristic is of particular significance as it characterizes the quality of synchronization in the case of practical application of the phenomenon of chaotic synchronization. In problems of information transmission with coherent reception, synchronization accuracy determines the accuracy of reproducing the received information.

Pull-out region of chaotic synchronization. This region in the space of parameters of dynamic models is the region of existence of chaotic synchronization. It is worth noting, however, that it is reasonable to consider the existence of the chaotic synchronization mode in connection with synchronization accuracy, as it determines the size of the pull-out region.

Pull-in region of chaotic synchronization. It is the region in the space of parameters where synchronization is guaranteed to set in under arbitrary or specific initial conditions for each concrete model. Hence, several approaches may be proposed to define the boundaries of the pull-in region. The boundary is defined either by the conditions of global stability of the manifold Υ^S, or by the condition of existence of a set of characteristic initial conditions for the attraction basin of Υ^S. The problem of defining the pull-in region of chaotic synchronization should be solved, like in the case of defining the pull-out region of chaotic synchronization, taking into consideration synchronization accuracy.

The onset time of synchronization mode. It is the time of the transient process from the initial state to the onset of chaotic synchronization, i.e., the time needed for the phase trajectory to reach an ϵ-neighborhood of the manifold Υ^S.

Study of the scenarios of the onset and loss of synchronization is of great importance in the investigation of chaotic synchronization, as understanding these mechanisms not only allows us in developing optimal algorithms of calculating dynamic characteristics of synchronization systems (pull-in and pull-out regions, synchronization accuracy, onset time of synchronization, and others), but also in minimizing errors in these calculations.

Let us now consider examples application of solution of the above problems in the model (6.1). The results of calculations of the boundaries of pull-in and pull-out regions of chaotic synchronization are presented in Fig. 6.4. Here, *curve* 1 defines the existence region of chaotic synchronization, and *curve* 2 bounds the region where this mode is globally stable, i.e., the pull-in region. The pull-out boundary is calculated by the following algorithm. The values of parameters and the initial states of PLL_1 and PLL_2 in the model (6.1) are

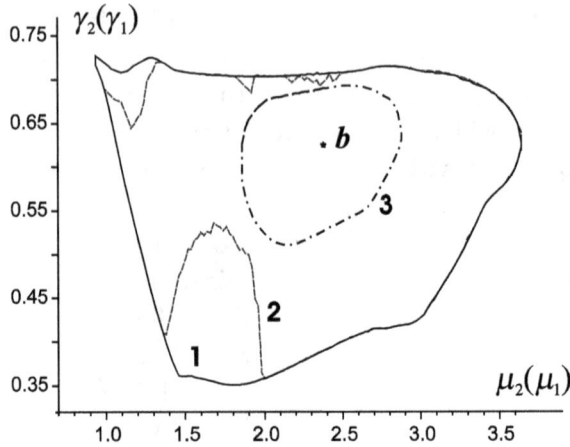

Fig. 6.4 Pull-in and pull-out regions of chaotic synchronization in the model (6.1) at $\varepsilon_1 = \varepsilon_2 = 1, \mu_1 = 2.38, \gamma_1 = 0.625, \delta_y = 0.5, \alpha = 30$: the point *b* denotes the values of PLL_1 parameters on the (μ_1, γ_1) plane, *curve* 1 is the boundary of the pull-out region of chaotic synchronization on the (μ_2, γ_2) plane (synchronization accuracy $\epsilon = 0.2$), *curve* 2 is the boundary of the pull-in region, and *curve* 3 is the boundary of the region of synchronization with enhanced accuracy (synchronization accuracy $\epsilon = 0.01$).

first taken to be identical. Chaotic synchronization sets in within the model (6.1).[1] Further, values of the parameters of the PLL_1 system remain unchanged, whereas values of PLL_2 parameters change, moving away from the values of PLL_1 parameters. The initial states of the model (6.1) for the new parameter values are inherited from the previous step. The mismatch between the PLL_2 and PLL_1 parameters increases until the synchronization is lost.

The boundary of the pull-in region is calculated in a reverse order. Now, values of parameters and the initial state of the model (6.1) are first chosen such that PLL_2 and PLL_1 should be in asynchronous mode. Further, the PLL_2 parameters change so that the mismatch

---

[1]Generally speaking, the initial states of PLL_1 and PLL_2 may be chosen to be non-identical, but in this case the initial conditions must belong to the attraction basin of the attractors defining the same dynamic mode. Then, for $\delta_y > 0$, chaotic synchronization will occur in a certain time interval T_{trans} whose duration depends on the ratio of the initial states of PLL_1 and PLL_2. The time T_{trans} is also finite; moreover, it is relatively small.

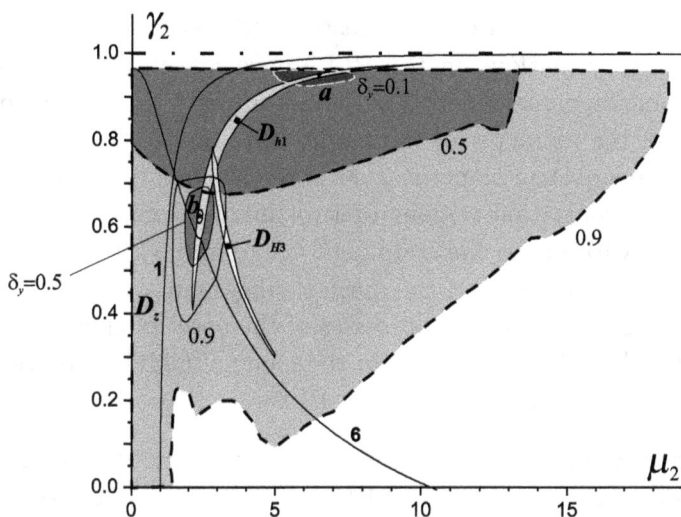

Fig. 6.5 Regions of existence of chaotic modulation within the model (6.1) for the case $\varepsilon_2 = 1$, $\delta_y = 0.1$; 0.5; 0.9, $\alpha = 30$ for the points $b\,(\gamma_1 = 0.625, \varepsilon_1 = 1, \mu_1 = 2.38)$ and $a\,(\gamma_1 = 0.95, \varepsilon_1 = 1, \mu_1 = 6.5)$. The boundaries of synchronization regions for the point b are shown by solid lines, and for the point a by dashed lines.

between the values of PLL$_1$ and PLL$_2$ parameters decreases; the initial states of the model (6.1) are inherited in this case. The parameter values at which chaotic synchronization sets in within the model (6.1) again correspond to the boundary of the pull-in region.

If we plot the boundaries of the pull-out regions of chaotic synchronization on the map of dynamic modes of the autonomous model of PLL$_2$ (Fig. 2.11(a)), we will obtain Fig. 6.5, from which it is clear that the synchronization regions of the model (6.1) may include the regions of existence of various dynamic modes of the autonomous model of PLL$_2$. From this, it follows that with an open control loop ($\delta_\varphi = \delta_y = \delta_z = 0$), PLL$_2$ can function nearly in all possible dynamic modes: from generation of regular oscillations synchronized with the reference signal to generation of CMO. Here, the control loops not only allow the compensation of a substantial difference between the two oscillatory processes, but also provide high accuracy of chaotic synchronization. Note that the onset of chaotic synchronization is appreciably affected by characteristics of the oscillatory motions as well as by parameters of the phase-locked loops. Figure 6.5 shows the chaotic oscillations in the PLL_1 have different amplitudes for the parameter values defined at points a and b, in particular, projections of the attractor of the model (6.1) on the (φ_1, y_1) plane at point a are less than at point b. As stated before, it follows that the CMO synchronization regions of two unidirectionally coupled systems grow with increasing coupling force δ_y and with decreasing size of the chaotic attractor of the master subsystem.

Let us now address the processes of the onset of chaotic synchronization, when PLL_1 functions in the mode of CMO generation, and different modes may be realized in PLL_2.

Example 1. Let PLL_1 parameters be defined at the point $b(\gamma_1 = 0.625, \varepsilon_1 = 1, \mu_1 = 2.38)$, and PLL_2 parameters have the following values: $\gamma_2 = 0.68, \varepsilon_2 = 1$, and $\mu_2 = 2$. Then in the absence of couplings $(\delta_\varphi = \delta_y = \delta_z = 0)$ in the partial phase space $U_1 = \{\varphi_1 \,(\text{mod}\, 2\pi), y_1, z_1\}$ in the model (6.1) there exists an oscillatory chaotic attractor, and in $U_2 = \{\varphi_2 \,(\text{mod}\, 2\pi), y_2, z_2\}$ there exist two regular attractors of different topologies — an oscillatory limit cycle L_1 (Fig. 6.6(a)) and a rotatory limit cycle L_2 (Fig. 6.7(a)). Let us define the PLL_2 state so that the cycle L_1 corresponding to a regular quasisynchronous mode is realized in the phase space U_2. Let us further increase values of the parameter δ_y attempting to synchronize the oscillations of PLL_1 and PLL_2. The process of the onset of synchronization is illustrated in Fig. 6.6 where one can see a one-parametric bifurcation diagram of a modified Poincaré map Fig. 6.6(b),[2] projections of phase projects and oscillograms Figs. 6.6(c–g). There is no synchronization at weak coupling $\delta_y = 0.05$ (Fig. 6.6(c)). As the coupling force is increasing up to $\delta_y = 0.15$, intervals of synchronous regimes rather frequently intermitting with asynchronous motions appear in the oscillogram (Fig. 6.6(d)). Such short intervals of synchronization do not markedly influence projections of the attractor phase portrait. A further increase in δ_y leads to

---

[2]The modified Poincaré map differs from a conventional one in that, instead of the coordinates of the Poincaré map, values of a certain function R of these coordinates are laid off on the y-axis on the diagram in Fig. 6.6(b). In particular, here, $R(y_2, y_1) = y_2 - y_1$.

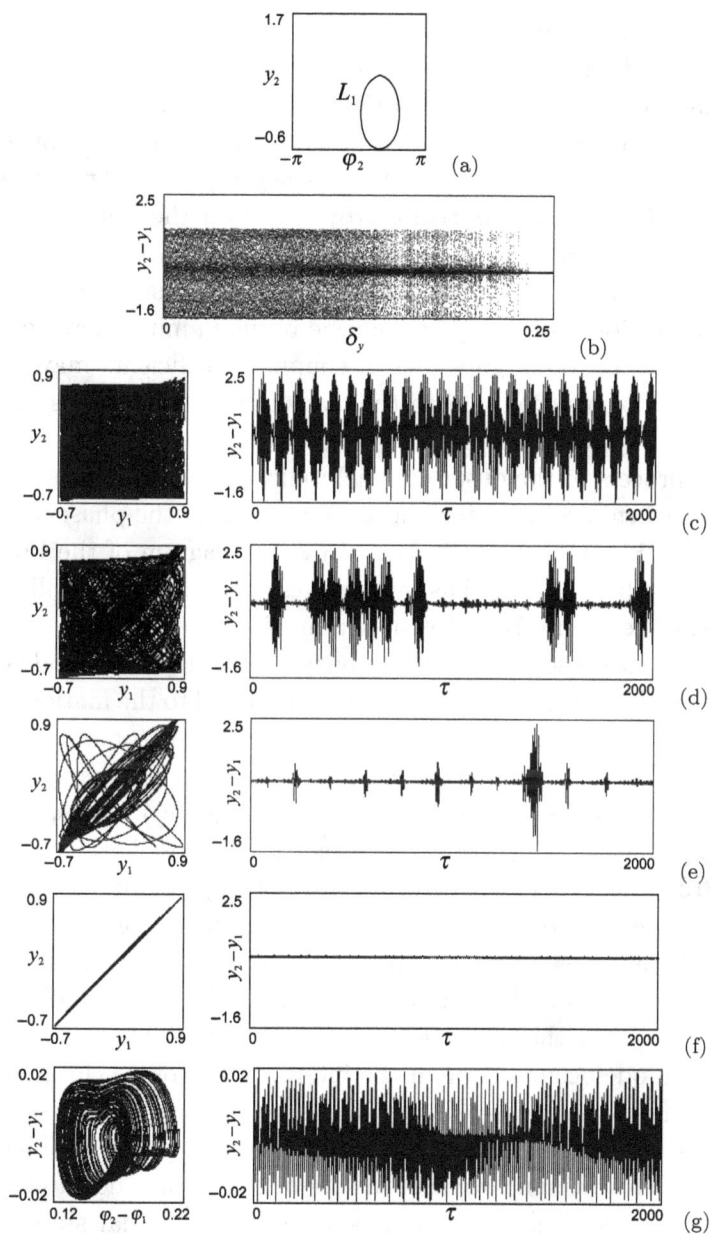

Fig. 6.6   Mode-locking of regular quasisynchronous oscillations of PLL$_2$ to CMO of PLL$_1$ for bistable behavior of PLL$_2$.

an increase in the intervals of synchronous motions and to the corresponding changes in the phase portrait projections (Fig. 6.6(e)) — the image point of the phase trajectories stays longer in the neighborhood of the diagonal. Finally, at $\delta_y = 0.232$, the synchronization mode with accuracy $\epsilon = 0.04$ in frequency $(|y_2(\tau) - y_1(\tau)| < \epsilon)$ sets in (Fig. 6.6(f)). Phase portrait projections on the difference plane of phase variables and the corresponding scaled-up oscillogram are shown in Fig. 6.6(g). A further increase in δ_y results in enhanced synchronization accuracy. It should be borne in mind, however, that introduction of too large values of coupling coefficients may lead to loss of synchronization and to the transition of the PLL$_2$ system to the mode of chaotic beats.

The process of the onset of synchronization with increasing δ_y in the case when a rotational cycle L_2 is realized in the phase space U_2 at $\delta_y = 0$ (Fig. 6.7(a)) is illustrated by the diagram of the Poincaré map in Fig. 6.7(b), as well as by the phase portraits and oscillograms in Figs. 6.7(c) and 6.7(d). Introduction of weak coupling $0 < \delta_y \ll 1$ leads to chaotic oscillations at the output of PLL$_2$. This mode corresponds to the phase space of the model (6.1) and to the motions along the attractor in which the difference between the φ_2 and φ_1 coordinates is constantly increasing and the difference between y_2 and y_1 changes chaotically in a wide interval (Fig. 6.7(c)). Thus, at weak coupling, there is no chaotic synchronization of oscillations in PLL$_1$ and PLL$_2$. The observed picture does not appreciably change with increasing δ_y up to $\delta_y = 0.234$ when the attractor of asynchronous modes breaks down and the phase trajectories from its neighborhood go to the synchronization manifold (Fig. 6.7(d)). If we now start to decrease the parameter δ_y, the transition from the synchronization mode occurs through the intermittence of synchronous and asynchronous motions at $\delta_y = 0.232$.

Thus, the considered example shows that in a system with bistable behavior, chaotic synchronization sets in and breaks down nearly at the same level of coupling, independent of the initial state of the model. The initial conditions determine the scenarios of the onset of the synchronization mode: for a regular quasisynchronous mode, the transition to the mode of CMO generation occurs softly through

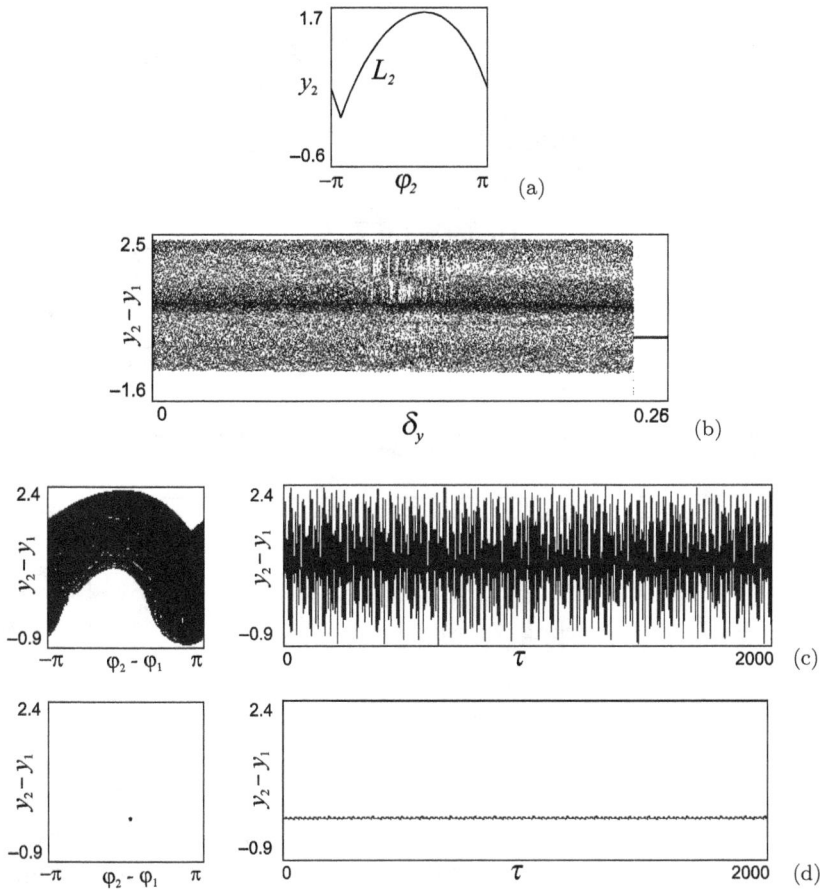

Fig. 6.7 Mode-locking of regular mode of beats of PLL$_2$ to CMO of PLL$_1$ for bistable behavior of PLL$_2$.

intermittence, and for regular beats the transition is hard through breakdown of an asynchronous attractor.

Example 2. Let PLL$_1$ parameters be defined at the point $b(\gamma_1 = 0.625$, $\varepsilon_1 = 1$, $\mu_1 = 2.38)$, and PLL$_2$ parameters have the following values: $\gamma_2 = 0.6$, $\varepsilon_2 = 1$, $\mu_2 = 2.7$. In the absence of couplings in the partial subspace U_1 of the model (6.1), there exists an oscillatory chaotic attractor, and in U_2 there exists a single attractor — a rotational limit cycle.

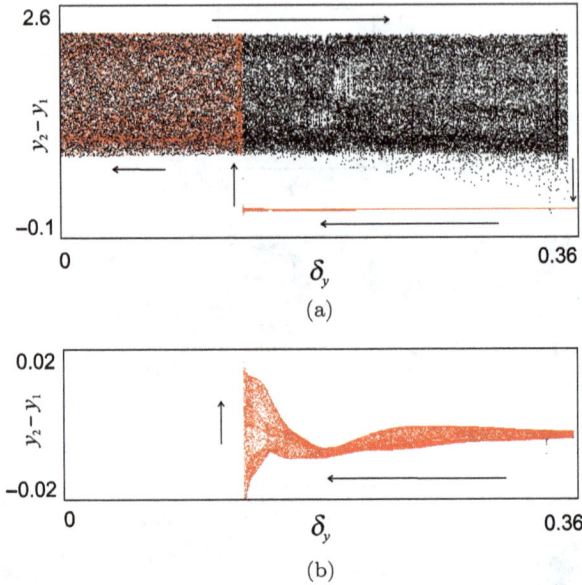

Fig. 6.8   Mode-locking of regular beats of PLL$_2$ to CMO of PLL$_1$.

The behavior of the oscillators of the ensemble as a function of the coupling force δ_y is illustrated in Fig. 6.8, where one can see a one-parametric bifurcation diagram of Poincaré map plotted for the values of δ_y increasing from 0 to 0.36 and further for δ_y decreasing to zero (a), as well as a magnified fragment of the Poincaré map representing changes at motion with decreasing coupling force.

Introduction of weak coupling, like in Example 1, leads to chaotization of oscillations at the PLL$_2$ output, but this coupling is insufficient to provide locking of PLL$_2$ oscillations to the chaotic oscillations of PLL$_1$. At weak couplings, the difference between the φ_2 and φ_1 coordinates is constantly increasing, and the y_2-y_1 difference changes in a wide region. There occurs a hard transition to the mode of chaotic synchronization when the level of coupling δ_y exceeds a certain critical value $(\delta_y)_2 = 0.355$. At $\delta_y = 0.36$, the frequencies of PLL$_1$ and PLL$_2$ oscillations coincide to an accuracy of $\epsilon = 0.02$. A further increase in coupling force results in enhanced synchronization accuracy, for example, at $\delta_y = 0.65$ the accuracy is $\epsilon = 0.01$, and at $\delta_y = 1.5$ the accuracy is $\epsilon = 0.004$. As the process of the onset of synchronization is hard, with increasing values of parameter δ_y the boundary of the pull-in region of synchronization does not coincide with the boundary of the pull-out region. In the considered example, loss of synchronization occurs from the synchronization manifold breakdown at $(\delta_y)_1 = 0.127$. This breakdown is preceded by a decrease in synchronization accuracy (Fig. 6.8(b)).

Thus, from Example 2, it follows that, even if originally (at $\delta_y = 0$) the behavior of two PLLs is defined unambiguously, the introduction of coupling aimed at achieving synchronization may lead to variable behavior of the ensemble. In the considered example, in the $(\delta_y)_1 < \delta_y < (\delta_y)_2$ interval, the model (6.1) demonstrates bistable behavior. Here, the synchronous and asynchronous modes coexist and the final result depends on initial conditions. It is clear from Fig. 6.5 that the size of the regions of chaotic synchronization depends on parameters of chaotic oscillations of the source (parameters of PLL_1 model) and control parameters. Values of PLL_1 parameters determine characteristics of the modulating oscillations. The region of variation of the phase variables φ_1 and y_1 on the attractors of the PLL_1 model characterize the modulation depth of the oscillations at the PLL_1 output, i.e., the smaller the attractor size, the lesser the modulation depth is. In the considered example, the size of the attractor determining the law of modulation for the parameter values of PLL_1 defined at point a is less than that at point b. From this as well as from the analysis of the curves in Fig. 6.5, it follows that the synchronization regions of CMO of two unidirectionally coupled systems grow with increasing coupling force δ_y and with decreasing size of the chaotic attractor of the master subsystem.

The possibility of synchronizing motions of different topologies strongly depends on control parameters. In the case under study, these parameters are δ_y and α. Figure 6.9 illustrates the relationship between the parameters of the PLL loop and the dynamic modes of autonomous PLL_2. The analysis of the presented results shows that quasisynchronous oscillations are the easiest to lock to CMO. Chaotic beat modes with oscillatory–rotatory behavior of trajectories are the worst to synchronize; the coupling force parameter δ_y for these oscillations to tune to CMO is much larger than for other types of oscillations.

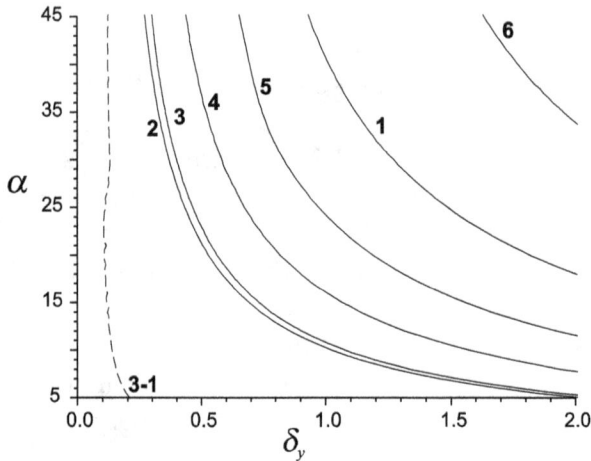

Fig. 6.9   Boundaries of pull-in regions of chaotic synchronization of oscillations of model (6.1) for different dynamic modes of autonomous PLL$_2$: *curve* 1 for synchronous mode at $\gamma_2 = 0.6, \varepsilon_2 = 1,\ \mu_2 = 1.0$; *curve* 2 for regular quasisynchronous mode at $\gamma_2 = 0.6, \varepsilon_2 = 1, \mu_2 = 2$; *curve* 3 for regular mode of beats at $\gamma_2 = 0.6, \varepsilon_2 = 1, \mu_2 = 2.7$; *curve* 4 for chaotic quasisynchronous mode at $\gamma_2 = 0.7, \varepsilon_2 = 1, \mu_2 = 2.63$; *curves* 5 and 6 for chaotic beats defined by rotational and oscillation–rotational chaotic attractors, respectively, at $\gamma_2 = 0.6, \varepsilon_2 = 1, \mu_2 = 3.2$ and $\gamma_2 = 0.69, \varepsilon_2 = 1, \mu_2 = 3.1$. The dashed line denotes the boundary of the pull-out region of chaotic synchronization corresponding to *curve* 3.

To conclude, the results of modeling system (6.1) demonstrate that PLL$_2$ oscillations can, in principle, be locked to (synchronized with) CMO in PLL$_1$ using the scheme presented in Fig. 6.1. In what follows, we will analyze the size of the regions of existence of the chaotic synchronization mode.

6.2.   Regions of Synchronization of CMOs

Let us set values of PLL$_1$ parameters in the region of existence of chaotic oscillations D_{H1} (Fig. 2.11): $\varepsilon_1 = 1, \mu_1 = 2.2, \gamma_1 = 0.5$; the initial values of φ_1, y_1, z_1 are specified on the oscillatory chaotic attractor that corresponds to the CMO mode. Then, we will search for values of parameters $\gamma_2, \varepsilon_2, \mu_2$ and of couplings $\delta_\varphi, \delta_y, a$ at which the CMO generated in PLL$_1$ and PLL$_2$ will differ only slightly.

The initial conditions in PLL_2 are set to be the following: $\varphi_2 \in [-\pi, \pi], y_2 = \gamma_2, z_2 = 0$. The values of φ_2 are calculated either as random quantities obtained by means of a random number generator or as coordinates of points equidistributed on the interval $-\pi < \varphi < \pi$. After the process sets in during time τ_y, we will calculate the deviations of the coordinates $\Delta\varphi = \varphi_1 - \varphi_2, \Delta y = y_2 - y_1, \Delta z = z_2 - z_1$ during the observation time τ_c. Further, we will analyze the deviations: If for a definite point in the space of parameters under all initial conditions of z_2, y_2, φ_2 (N_φ points on the interval $[-\pi, \pi]$), the deviations $\Delta\varphi, \Delta y, \Delta z$ do not exceed the limits of the specified accuracy ε_c during time τ_c, then this point belongs to the synchronization region. In view of the chaotic character of the calculated trajectories, the observed coordinates are allowed to go outside the specified accuracy ε_c for a short time, so that the total time outside this accuracy should be less than 10% of τ_c. The sections of the synchronization region (δ_y, γ_2), (μ_2, γ_2), and $(\varepsilon_2, \gamma_2)$ plotted by this algorithm at $N_\varphi = 30, \epsilon_c = 0.025, \tau_y = 1000, \tau_c = 1000$ are presented in Figs. 6.10(a–c), respectively. It is clear from this figure that synchronization is, indeed, realized in a rather wide region of parameters, with the values of PLL_2 parameters not necessarily belonging to the region D_{H1} (Fig. 2.11). Note that the region of synchronization in Fig. 6.10(a) must evidently be bounded at rather large δ_y, but this boundary has not been calculated here.

The synchronization of regular oscillations is known to be a hysteretic process, i.e., pull-in and pull-out of synchronization depend on the value of oscillator initial frequency detuning relative to the reference oscillator. It would be the interesting to find whether such hysteresis is typical for synchronization of chaotic oscillations [43].

For synchronization regions considered above (Fig. 6.10), we conducted experiments aimed at calculating the boundaries of synchronization regions with a slow increase (pull-out) and slow decrease (pull-in) of parameter γ_2. It was found that pull-out and pull-in of synchronization occur at the same value of γ_2, which means that there is no hysteresis ($\gamma_2^{\text{pull-in}} = \gamma_2^{\text{pull-out}}$). This is explained by the fact that, for the chosen values of parameters, the chaotic attractor is the only attractor of PLL_1 model in region D_{H1}. However,

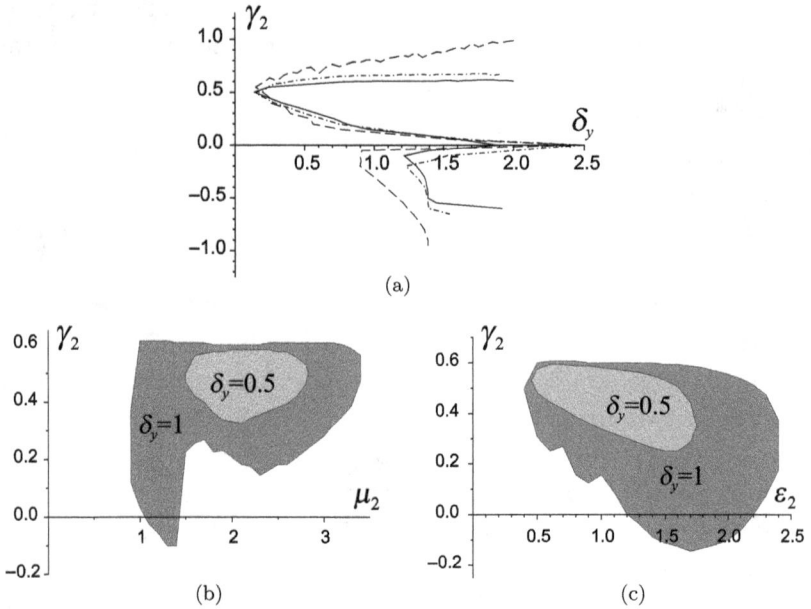

Fig. 6.10 Regions of CMO synchronization of the model (6.1) for $\varepsilon_2 = 1, \mu_2 = 2.2, \delta_\varphi = 0$ — solid curve; $\delta_\varphi = 0.1$ — dash-dot curve; $\delta_\varphi = 0.5$ — dashed curve (a); (b) $\varepsilon_2 = 1, \delta_\varphi = 0, \delta_y = 0.5$ and 1 — light and dark regions; (c) for $\mu_2 = 2.2, \delta_\varphi = 0, \delta_y = 0.5$ and 1 — light and dark regions.

hysteresis may be realized ($\gamma_2^{\text{pull-in}} \neq \gamma_2^{\text{pull-out}}$) for other values of parameters from this region. Such an example is given in Fig. 6.11 for the following values of parameters PLL$_1$: $\gamma_1 = 0.7$, $\varepsilon_1 = 1$, $\mu_1 = 2.6$; PLL$_2$: $\varepsilon_2 = 1.4$, $\mu_2 = 2.5$, and parameters of the control circuit: $\delta_y = 0.5$, $\delta_\varphi = 0$, $a = 20$. Parameter γ_2 increases from 0.7 to 0.8, and then decreases from 0.8 to 0.7. A one-parametric bifurcation diagram $\{\gamma_2, y_2\}$ of the Poincaré map depicted in Fig. 6.11(a) shows that as γ_2 changes from 0.7 to 0.8, first chaotic oscillations synchronized with oscillations in PLL$_1$ (Fig. 6.11(c)) are present at the output of PLL$_2$ (Fig. 6.11(b)). Then, at $\gamma_2^{\text{pull-out}} = 0.77$ (the boundary of pull-out region) the synchronization is lost and further, with a decrease in γ_2 (the top of Fig. 6.11(a)), PLL$_2$ chaotic oscillations (Fig. 6.11(d)) are no longer synchronized with PLL$_1$ oscillations (Fig. 6.11(e)). The subsequent pulling into synchronization of chaotic oscillations occurs at $\gamma_2^{\text{pull-in}} = 0.703$.

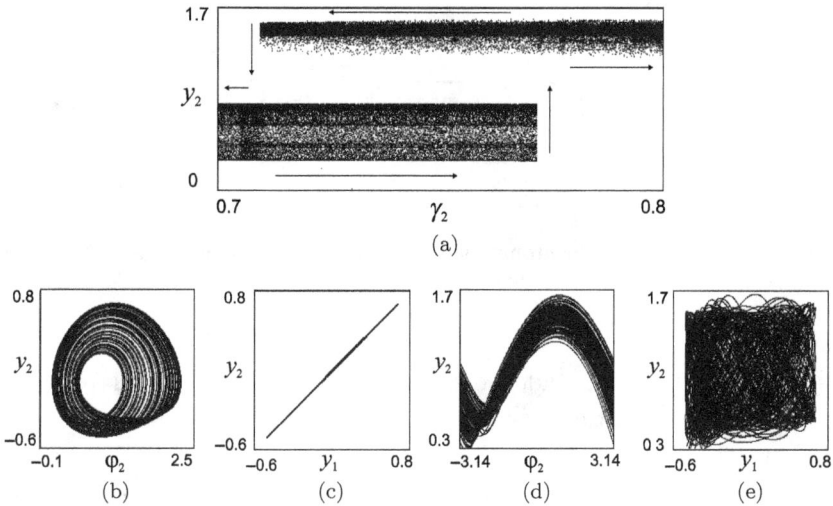

Fig. 6.11   Hysteresis at CMO synchronization.

Thus, the above consideration enables us to make the conclusion that CMO may not only be generated but also synchronized in a PLL system within a rather wide region of parameter variation. Processes of CMO synchronization may be hysteretic, i.e., the borders of pull-in and pull-out regions do not coincide within the space of parameters.

In spite of the fact that computer simulation confirmed the possibility of attaining the mode of complete CMO synchronization in two non-identical PLLs with close values of parameters, it will be difficult to implement such synchronization of two chaotic PLLs with second-order filters in practice because of relative smallness of the existence regions of oscillatory chaotic attractors within the space of parameters.

6.3.   Synchronization of CMOs in Ensembles

We now move to the problem of synchronization of chaotic oscillations generated by an ensemble of cascade-coupled PLL_1 and PLL_2 further referred to as $PLL_1 + PLL_2$ for brevity [80]. There are different variants of solving this problem. One of the obvious solutions is the following. Let $PLL_1 + PLL_2$ be a master chaotic oscillator and

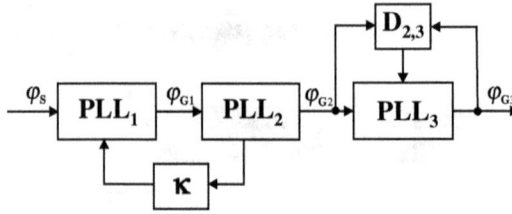

Fig. 6.12   Scheme of synchronization of chaotic oscillations of cascade-coupled oscillators G2 and G3.

the $\text{PLL}_3 + \text{PLL}_4$ cascade be a synchronized chaotic oscillator with close but not identical parameters. To tune the oscillations at the output of PLL_4 to the oscillations at the output of PLL_2, i.e., to achieve synchronization, unidirectional difference coupling is introduced between the master oscillator and the synchronized cascade, like it was done for the PLL in Fig. 6.1. However, taking into consideration specific features of cascade coupling, we can suggest a simpler solution.

In Fig. 6.12, we present a variant of cascade coupling of three PLLs that gives chaotic oscillations at the output of G3 which are synchronous with the oscillations at the output of G2. The synchronization is achieved by the principle of synchronous response [19]. In this case, the master cascade $\text{PLL}_1 + \text{PLL}_2$ is cascade coupled with PLL_3. If PLL_3 has a low-inertia filter with transmission coefficient $K(p) = 1/(1 + T_3 p)$, then chaotic oscillations stimulated by chaotic oscillations at the input (oscillations from PLL_2 output) will arrive at the output of PLL_3. For tuning G3 oscillations to G2 oscillations, control coupling from PLL_2 to PLL_3 is introduced through frequency discriminator FD_{23}. Such a system is described by the following equations:

$$\frac{d\varphi_1}{d\tau} = \gamma_1 - \sin\varphi_1 - \kappa\sin(\varphi_2 - \varphi_1),$$

$$\frac{d\varphi_2}{d\tau} = y_2, \quad \varepsilon_2\frac{dy_2}{d\tau} = \gamma_2 - y_2 - \sin(\varphi_2 - \varphi_1), \tag{6.4}$$

$$\frac{d\varphi_3}{d\tau} = y_3, \quad \varepsilon_3\frac{dy_3}{d\tau} = \gamma_3 - y_3 - \sin(\varphi_3 - \varphi_2) - \delta_y\Phi(y_3 - y_2),$$

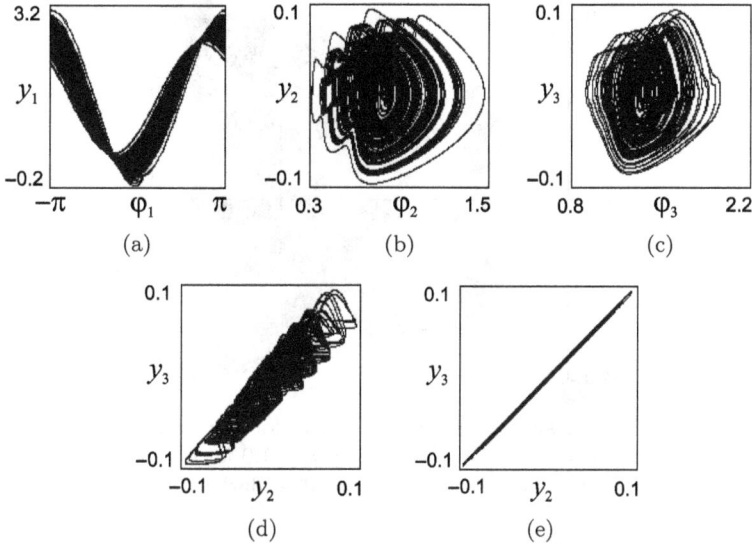

Fig. 6.13   Oscillations at the output of (a) G1, (b) G2, (c) G3 and results of their synchronization in the absence (d) and (e) in the presence of control.

where $\Phi(y_3 - y_2)$ is the nonlinear characteristic of FD$_{23}$. Modeling of the system (6.4) for $\gamma_1 = 15$, $\varepsilon_1 = 0$, $\gamma_2 = 0.72$, $\varepsilon_2 = 130$, $\gamma_3 = 0.5$, $\varepsilon_3 = 5$, $\kappa = 1.9$ showed that for initial conditions $y_2^0 = \gamma_2$, $y_3^0 = \gamma_3$, φ_2^0 and φ_3^0 being arbitrary values and zero control coupling $\delta_y = 0$, the chaotic oscillations of G2 and G3 are not synchronized (Figs. 6.13(a–d)). However, introduction of control coupling $\delta_y = 0.8$, $a = 20$ (nonlinearity $\Phi(y_3 - y_2) = 2a(y_3 - y_2)/[1 + a^2(y_3 - y_2)^2]$) allows G3 oscillations to be tuned to G2 oscillations and good enough synchronization is attained (Fig. 6.13(e)).

CMO synchronization regions for the system (6.4) on the plane of parameters (δ_y, γ_3) are depicted in Fig. 6.14. These regions were obtained by the algorithm used above for the calculation of CMO synchronization regions of a PLL.

Still another, more efficient variant of solution to the problem of synchronization using parallel connection of PLL$_3$ to the PLL$_1$ + PLL$_2$ cascade is proposed in Fig. 6.15. Here, PLL$_3$ is a complete analog of PLL$_2$ and the input signal from PLL$_1$ is fed both to the input of PLL$_2$ and to the input of PLL$_3$. Consequently, for identical

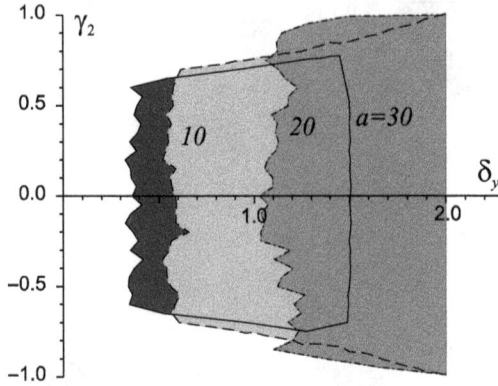

Fig. 6.14 Regions of synchronization of CMO for the scheme in Fig. 6.12 at $\gamma_1 = 1.7$, $\varepsilon_1 = 0$, $\gamma_2 = 0.72$, $\varepsilon_2 = 150$, $\varepsilon_3 = 5$, $\kappa = 1.9$ for $a = 30$ — the border is a solid curve, $a = 20$ — dash-dot curve, $a = 10$ — dashed curve.

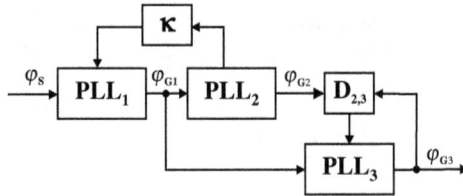

Fig. 6.15 Scheme of synchronization of chaotic oscillations of G2 and G3 coupled in parallel.

PLL_2 and PLL_3, chaotic oscillations formed at the output of PLL_3 will be the same as at the output of PLL_2.

For parameter deviations in PLL_2 and PLL_3, the oscillations of G3 should be tuned to the oscillations of G2 by coupling the two oscillators through the frequency discriminator FD_{23}. Equations for such a system may be written in the form

$$\frac{d\varphi_1}{d\tau} = \gamma_1 - \sin\varphi_1 - \kappa\sin(\varphi_2 - \varphi_1),$$

$$\frac{d\varphi_2}{d\tau} = y_2, \quad \varepsilon_2\frac{dy_2}{d\tau} = \gamma_2 - y_2 - \sin(\varphi_2 - \varphi_1), \qquad (6.5)$$

$$\frac{d\varphi_3}{d\tau} = y_3, \quad \varepsilon_3\frac{dy_3}{d\tau} = \gamma_3 - y_3 - \sin(\varphi_3 - \varphi_1) - \delta_y\Phi(y_3 - y_2).$$

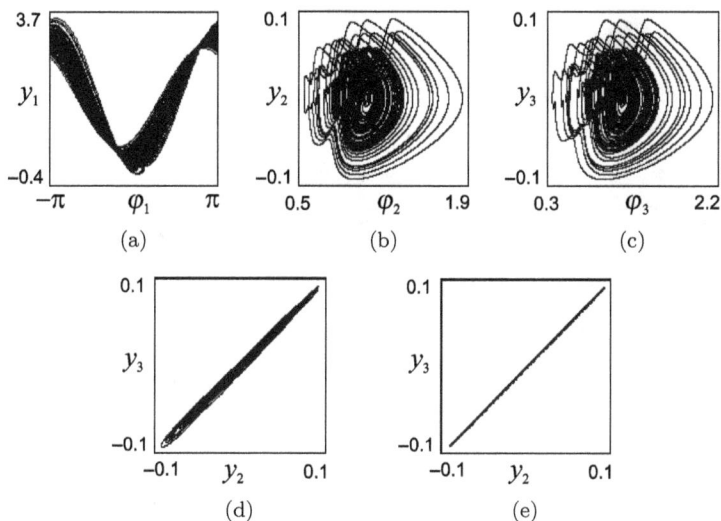

Fig. 6.16 Oscillations at the output of (a) G1, (b) G2, (c) G3 and results of their synchronization with (d) linear and (e) nonlinear control.

System (6.5) was modeled under initial conditions $y_2^0 = \gamma_2$, $y_3^0 = \gamma_3$ and arbitrary φ_2^0 and φ_3^0. Deviations of the parameters $\gamma_2 - \gamma_3 = \Delta\gamma \neq 0$, $\varepsilon_2 - \varepsilon_3 \neq 0$ were specified within 5%. The results of modeling verified that by using frequency coupling it is possible to tune G3 chaotic oscillations to G2 oscillations in a rather wide region of parameter variation (Fig. 6.16). The synchronous mode is illustrated in Fig. 6.16(d) (with linear control coupling) and in Fig. 6.16(e) (with nonlinear control coupling).

To conclude, we have found that by using small PLL ensembles, it is possible, in principle, to achieve efficient generation and synchronization of CMO. As traditional PLL systems (with regular oscillations) are broadly employed for modulation and demodulation of oscillations (signal coding and decoding), they look promising for solving a whole complex of problems needed for information communication by using a coherent reception of chaotically modulated signals.

Chapter 7

Communication with Chaos

The results obtained on the generation and synchronization of chaotically modulated oscillations (CMO) may lay a basis for different variants of information communication using them as carrier oscillations [80].

7.1. Transmission of a Binary Signal

Consider a variant for constructing a system for binary signal transmission. Using the method of chaos shift keying [81], it is possible to design a transmitter comprising two chaotically modulated oscillators (CMO$_1$ and CMO$_2$) keyed in line with the specified binary information signal and use CMO$_3$ oscillator in the receiver with parameters such that it may be synchronized only with one oscillator of the transmitter, for example, CMO$_1$. A block diagram of such a transmitter of information using chaotic signal keying is drawn in Fig. 7.1. Information message S in the form of a binary signal controls the operation of the transmitter switch. Let the arrival of binary zero (**0**) indicate the transmission of a chaotically modulated signal x_1 from the first oscillator, and of binary unit (**1**) of signal x_2 from the second oscillator. In this case, a chaotic signal formed in conformity with the information signal from CMO$_1$ and CMO$_2$ will always be present at the transmitter output. As CMO$_3$ oscillator is synchronized with the signal of the CMO$_1$ oscillator only, the signal at the receiver output on subtraction will be different from zero only in the absence of synchronization, i.e., it corresponds to the reception of binary unit (**1**).

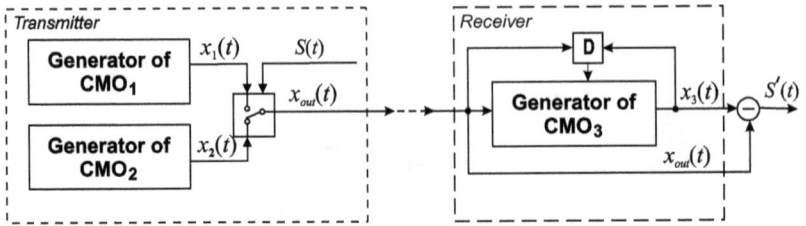

Fig. 7.1   Block diagram of information transmission using chaotic signal keying.

Figure 7.2 shows the results of computer simulation of binary signal transmission illustrating the time variation of information signal S (Fig. 7.2(b)), signal at the transmitter output (Fig. 7.2(c)), signal S' at the receiver output (Fig. 7.2(d)), and projections of the attractors corresponding to CMO_1, CMO_2 and CMO_3 (Fig. 7.2(a)). Of course, there may be other variants for realizing the idea of the considered approach.

7.2. Transmission of a Continuous Signal

Let us now consider the transmission of an analog signal [59]. A scheme of information communication in which chaotic oscillations generated by cascade-coupled phase-locked loops (PLLs) are used as carrier ones is demonstrated in Fig. 7.3. The transmitter consists of three PLL systems. The $PLL_1 + PLL_2$ cascade generates a carrier — CMOs with central frequency stabilized by the reference signal. PLL_3 acts as a modulator; the chaotic modulation of the carrier at its output is supplemented by modulation by the information signal. Parameters of PLL_3 must be close to those of PLL_2. The G3 oscillations are tuned to G2 oscillations through control discriminator FD_{23}. Signals from the outputs of G1, G2 and G3 are transmitted to the receiver through three communication channels. A signal from the output of G3 in the receiver is transmitted to PLL_5 demodulator whose voltage is proportional to the mixture of the chaotically modulated carrier signal and the information signal. Signals from the outputs of G1 and G2 arrive at PLL_4 for the reconstruction of a copy of the chaotically modulated carrier signal which

(a)

(b)

(c)

(d)

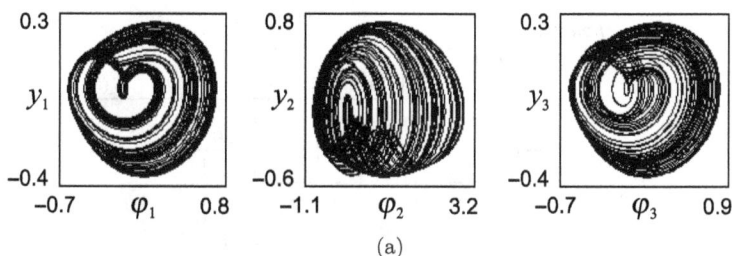

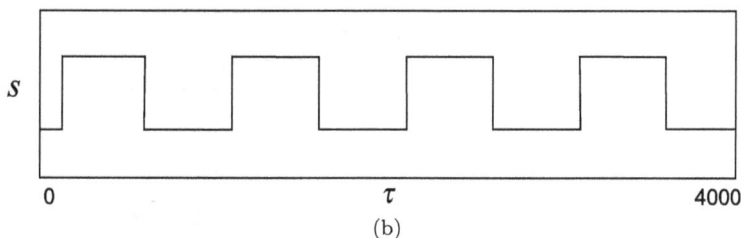

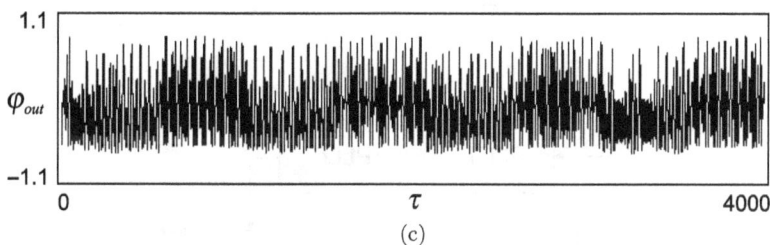

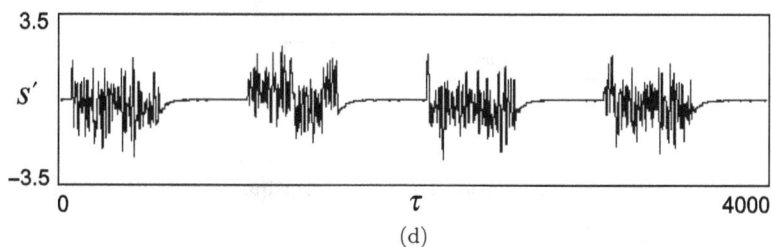

Fig. 7.2   Simulation process of binary signal transmission.

is then demodulated in PLL_6. The voltage at the output of PLL_6 demodulator is proportional to the chaotic modulation of the carrier. A useful information signal is obtained after subtraction of the signals from PLL_5 and PLL_6 demodulators. Comparison of the received and transmitted signals obtained in computer simulation demonstrates

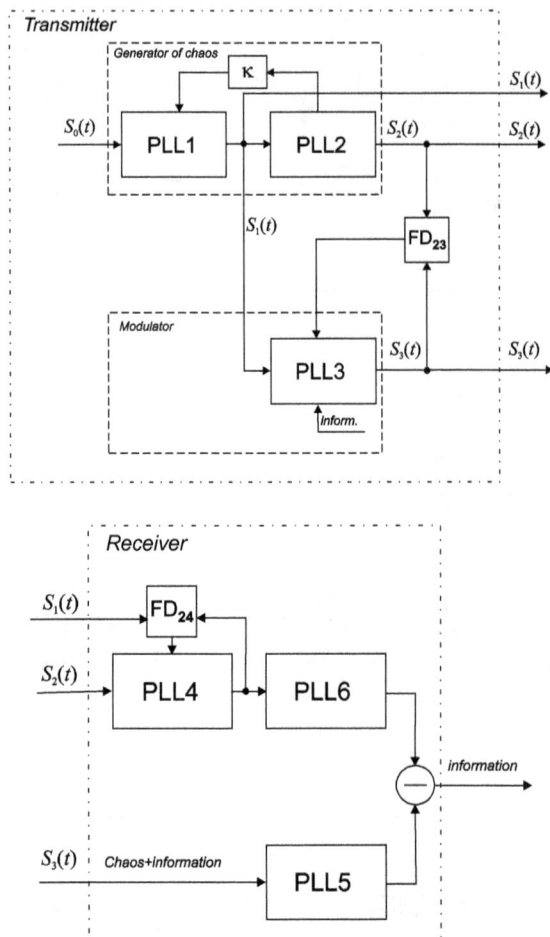

Fig. 7.3   Scheme of information transmission using cascade-coupled PLLs.

that the considered scheme ensures a good enough quality of signal transmission, even when PLL systems in the receiver and transmitter are significantly non-identical.

Results of the simulation of information communication system with parameters of PLL_3 and PLL_4 set to deviate from parameters of PLL_2 within 5% are illustrated in Fig. 7.4. In this figure, one can see oscillograms at the output of phase discriminators of PLL_1–PLL_4 (Figs. 7.4(a–d)) from PLL_5 and PLL_6 demodulators (Figs. 7.4(e)

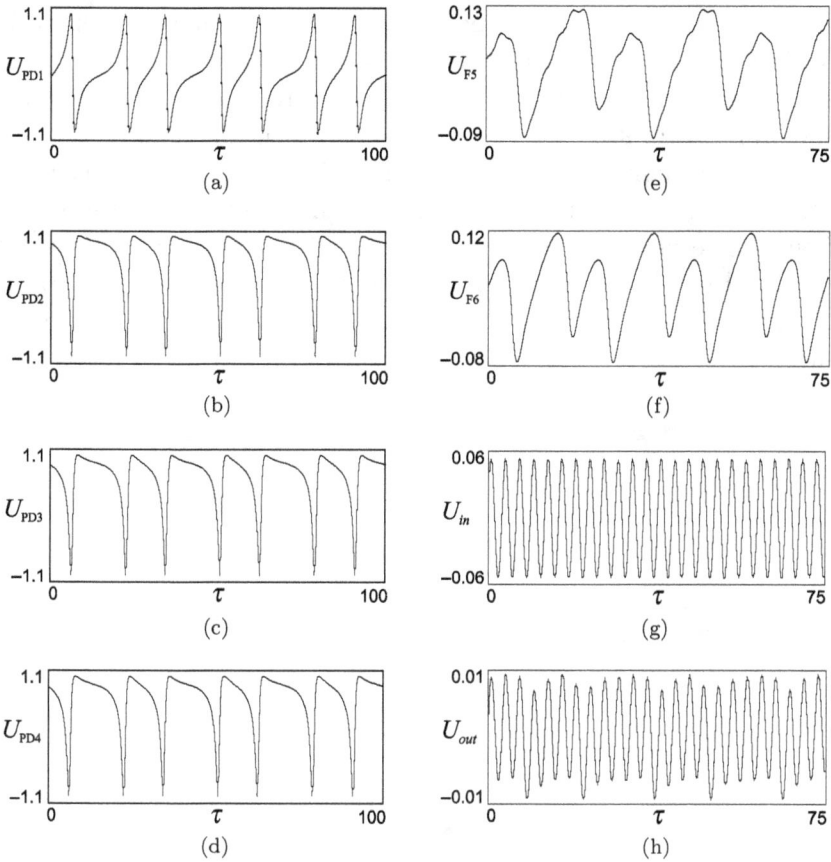

Fig. 7.4   Results of simulation of analog signal transmission.

and 7.4(f)) of the transmitted signal (Fig. 7.4(g)) and the received (Fig. 7.4(h)) signal, respectively.

The presented results of simulation lead to the conclusion that a system of information communication using chaotically modulated oscillations may, in principle, be constructed based on collective dynamics of an ensemble of coupled PLLs (cascade-coupled PLLs in particular). Today, we understand that collective dynamics of PLL systems provide wide opportunities for reliable generation of various chaotic oscillations. Synchronization, i.e., tuning of chaotically modulated oscillations, may be attained by additional couplings of

PLLs, for which, generally speaking, three communication channels between the transmitter and the receiver are needed. However, more stringent conditions on the deviation of parameters of the receiver and transmitter systems will allow reduction in the number of communication channels. For instance, the third communication channel may be rejected. In this case, PLL_4 is unnecessary and the signal from the second channel must be transmitted in a straightforward manner to PLL_6 demodulator. It is also obvious that for the transmission of binary signals, one communication channel will be quite sufficient.

Chapter 8

Conclusion

Needless to say, the presented brief survey of the features of chaotic oscillations in ensembles of coupled phase-controlled oscillators is not exhaustive. Nevertheless, it demonstrates quite convincingly that phase-locked loop (PLL) systems may be successfully used for generation and synchronization of chaotically modulated oscillations and are, thus, promising for the development of novel systems of information communication with coherent reception. There naturally arise many other important problems to be solved that are beyond the scope of the current work, for example, noise stability of such systems. Note that in traditional communication systems, where regular signals are employed, coherent receivers based on PLL possess optimal filtering properties [30]. This gives us grounds to expect that PLL communication systems with chaotic signals will provide acceptable noise stability.

We have already mentioned that the pioneering, although not too successful, attempts to solve the problem of information transmission using dynamical chaos based on PLL systems [21,82,83] were made at the beginning of the 1990s but did not meet with the support of researchers in the years to follow. This may be attributed to the fact that in those works, chaos was intended to be generated in a single PLL system but no comprehensive analysis of chaotic modes in a single PLL was available in the literature at that time. In particular, chaotic oscillations were not classified. Detailed analysis of chaotic modes in such systems undertaken later [33, 40, 41] verified that it is, as a rule, impossible to attain reliable generation of chaos (chaotically modulated oscillations [CMO]). Regions of CMO in a single

PLL are relatively small, hence even small uncontrolled parameter variation may result in the transition of the system from this domain. Consequently, there arose the idea to use small ensembles of coupled PLLs for CMO generation [26, 59].

Toward further development of this idea presented in the book, the current work has demonstrated that the use of the properties of collective dynamics of small PLL ensembles provides a whole set of variants for connecting PLL systems in an ensemble for producing an effective source of various chaotic oscillations (chaotically modulated oscillations, chaotic beats).

Today, as industries have mastered the manufacturing of integrated PLL circuits for different frequency ranges, the design and fabrication of integrated circuits of ensemble of coupled PLLs as chaotic oscillators seem quite feasible.

For communication, it is reasonable to choose chaotically modulated oscillations with central frequency that are stabilized by the reference frequency. Although spectral characteristics of CMO (spectrum width, spectrum diversity) are frequently inferior to those of other types of chaotic oscillations generated by PLL ensembles (chaotic beats, mixed chaotic oscillations), the chaotically modulated oscillations provide reliable synchronization. It is important to note that synchronization may be implemented in a rather wide interval of parameter variation, i.e., generators of chaos in the receiver and the transmitter may be significantly non-identical, even if they have close parameters.

The presented results were obtained primarily by computer simulations, although we also present some data of experiments on chaos generation. Naturally, further development of the considered ideas should rely on wide experimental testing of novel communication circuits using chaotically modulated oscillations, which will further allow us to pass over from model situations to real specific tasks of communication and practical engineering design.

Bibliography

[1] R. N. Madan (ed.) *Chua's Circuit: A Paradigm for Chaos*, World Scientific Publishing Co., Singapore, 1993.

[2] M. I. Rabinovich, P. Varona, A. I. Selverston and H. D. I. Abarbanel, "Dynamical principles in neuroscience," *Rev. Mod. Phys.*, Vol. 78, p. 1213, 2006.

[3] N. J. McCullen, M. V. Ivanchenco, V. D. Shalfeev and W. F. Gale, "A dynamical model of decision-making behaviour in a network of consumers with applications to energy choices," *Int. J. Bifurc. Chaos*, Vol. 21, pp. 1–14, 2011.

[4] V. S. Afraimovich, V. I. Nekorkin, G. V. Osipov and V. D. Shalfeev, *Stability, Structures and Chaos in Nonlinear Synchronization Networks*, World Scientific Publishing Co., Singapore, 1994.

[5] A. J. Lichtenberg and M. A. Lieberman, *Regular and Stochastic Motions*, Springer, Berlin, 1983.

[6] M. I. Rabinovich and D. I. Trubetskov, *Introduction to the Theory of Oscillations and Waves*, Kluwer Academic Publishers, Amsterdam, 1989.

[7] L. O. Chua (ed.), "Chaotic system," *Proc. IEEE*, Vol. 75, No. 8, 1987.

[8] G. S. Shuster *Deterministic Chaos: An Introduction*, Physik-Verlag, Weinheim, 1984.

[9] A. S. Dmitriev and V. Y. Kislov, *Stochastic Oscillations in Radiophysic and Electronic Circuits*, Nauka, Moscow, 1989 (in Russian).

[10] F. C. Moon, *Chaotic Vibrations: An Introduction for Applied Scientists and Engineers*, John Wiley & Sons, 1990.

[11] P. Berge, Y. Pomeau and C. Vidal, *L'ordre dans le Chaos: Vers une approche déterministe de la turbulence*, Nouvelle édition corrigée, Paris, 1988.

[12] A. S. Dmitriev and A. I. Panas, *Dynamic Chaos: Novel Type of Information Carrier for Communication System*, Izd. Fizmat. Lit. Moscow, 2002 (in Russian).

[13] V. S. Anishchenko, T. E. Vadivasova and V. V. Astakhov, *Nonlinear Dynamics of Chaotic and Stochastic Systems*, Saratov University Publication, 1999 (in Russian).

[14] H. Fujisaka and T. Yamada, "Stability theory of synchronized motion in coupled oscillator systems," *Prog. Theor. Phys.*, Vol. 69, No 1, pp. 32–46, 1983.

[15] A. S. Pikovsky, "On the interaction of strange attractors," *Z. Physik B.*, Vol. 55, p. 149, 1984.

[16] V. S. Afraimovich, N. N. Verichev and M. I. Rabinovich, "Stochastic synchronization of oscillation in dissipative systems," *Radiophys. Quantum Electron.*, Vol. 29, p. 795, 1986.

[17] L. M. Pecora and T. L. Caroll, "Synchronization of chaotic systems," *Chaos*, Vol. 25, No. 9, p. 097611, 2015.

[18] D. M. Abrams, L. M. Pecora and A. E. Motter, "Introduction to focus issue: Patterns of network synchronization," *Chaos*, Vol. 26, No. 9, p. 094601, 2016.

[19] A. R. Volkovskii and N. F. Rul'kov, "Synchronous chaotic response of a nonlinear oscillator system as a principle for detection of the information component of chaos," *Tech. Phys. Lett.*, Vol. 19, No. 2, pp. 97–99, 1993.

[20] L. Kocarev, K. S. Halle, K. Eckert, L. O. Chua and U. Parlitz, "Experimental demonstration of secure communication via chaotic synchronization," *Int. J. Bifurc. Chaos*, Vol. 2, No. 3, pp. 705–713, 1992.

[21] A. S. Dmitriev, A. I. Panas and S. O. Starkov, "Experiments on speech and music signals transmission using chaos," *Int. J. Bifurc. and Chaos*, Vol. 5, No. 4, pp. 1249–1254, 1995.

[22] A. S. Dmitriev, L. V. Kuz'min, A. I. Panas and S. O. Starkov "Experiments on the data transmission over a radio channel using chaos," *J. Commun. Technol. Electron.*, Vol. 43, No. 9, p. 1038, 1998.

[23] G. Kolumban, M. P. Kennedy and L. O. Chua, "The role of synchronization in digital communications using chaos. I. Fundamentals of digital communications," *IEEE Trans. Circuits Syst. I, Fundam. Theory Appl.*, Vol. 44, No. 10, pp. 927–936, 1997.

[24] G. Kolumban, M. P. Kennedy and L. O. Chua, "The role of synchronization in digital communications using chaos. II. Chaotic modulation and chaotic synchronization," *IEEE Trans. Circuits Syst. I, Fundam. Theory Appl.*, Vol. 45, pp. 1129–1140, 1998.

[25] G. Kolumban and M. P. Kennedy, "The role of synchronization in digital communications using chaos. III. Performance bounds for correlation receivers," *IEEE Trans. Circuits Syst. I, Fundament. Theory Appl.*, Vol. 47, No. 12, pp. 1673–1683, 2000.

[26] V. Shalfeev, V. Matrosov and M. Korzinova, "Dynamical chaos in ensembles of coupled phase systems," *Zarubezhnaja Radio Electronika, Uspekhi Sovremennoi Radioelektroniki*, No. 11, pp. 44–56, 1998 (in Russian).

[27] A. J. Viterbi, *Principles of Coherent Communications*, McGraw-Hill Book Company, New York, 1966.

[28] V. V. Shakhgildyan and A. A. Lyahovkin, *Phase Lock Loops*, Svyaz, Moscow, 1972 (in Russian).

[29] V. V. Shakhgildyan and L. N. Belyustina (eds.), *Phase Synchronization*, Svyaz, Moscow, 1975 (in Russian).

[30] W. Lindsey, *Synchronization Systems in Communication and Control*, Prentice-Hall, Englewood Cliffs, NJ, 1972.

[31] V. V. Shakhgildyan and L. N. Belyustina (eds.), *Phase Synchronization Systems*, Radio i Svyaz, Moscow, 1982 (in Russian).

[32] M. V. Kapranov, V. N. Kuleshov and G. M. Utkin, *Theory of oscillations in radioengineering*, Nauka, Moscow, 1984 (in Russian).

[33] V. V. Matrosov, "Regular and chaotic oscillations of phase systems", *Techn. Phys. Lett.*, Vol. 22, No. 23, pp. 4–8, 1996.

[34] M. V. Kapranov, "Pull-in band in PLL," *Radiotechnika*, Vol. 11, No. 12, pp. 37–52, 1956 (in Russian).

[35] L. N. Belyustina, "Dynamic of PLL," *J. Radiophys. Quantum Electron.*, Vol. 2, No. 2, pp. 277–291, 1959.

[36] L. N. Belyustina and V. N. Belykh, "Dynamic system on cylinder," *Diff. Uravneniya*, Vol. 9, No. 3, pp. 403–415, 1973 (in Russian).

[37] L. N. Belyustina, V. V. Bykov, K. G. Kiveleva and V. D. Shalfeev, "Pull-in range of PLL with proportionally integrating filter," *J. Radiophys. Quantum Electron.*, Vol. 13, No. 4, pp. 561–566, 1970.

[38] A. A. Andronov, A. A. Witt and S. E. Khaykin, *Theory of Oscillators*, Gostekhizdat, Moscow 1937 (in Russian); Pergamon Press, New York, 1966.

[39] V. N. Belykh and V. I. Nekorkin, "Bifurcations in the third-order nonlinear equation of phase synchronization," *Prikladnaya Matematika*, Vol. 42, No. 5, pp. 808–815, 1978 (in Russian).

[40] V. V. Matrosov, "Nonlinear dynamics of phase-locked loop with the second-order filter," *Radiophys. Quantum Electron.*, Vol. 49, No. 3, pp. 239–249, 2006.

[41] V. V. Matrosov, "Self-modulation regimes of a phase-locked loop with the second-order filter," *Radiophys. Quantum Electron.*, Vol. 49, No. 4, pp. 322–332, 2006.

[42] V. V. Matrosov, "Regular and chaotic oscillations of phase-locked loop with the second-order filter," *Proc. NDES'97*, Moscow, Russia, pp. 554–558, 1997.

[43] V. D. Shalfeev and V. V. Matrosov, "Pull-in and pull-out effects of synchronization of chaotic modulated oscillations," *Radiophys. Quantum Electron.*, Vol. 41, No. 12, pp. 1033–1036, 1998.

[44] V. D. Shalfeev, V. V. Matrosov and M. V. Korzinova, "Chaos in phase systems: generation and synchronization," *Controlling Chaos and Bifurcations in Engineering Systems*, ed. G. Chen, CRC Press, Boca Raton, pp. 529–558, 1999.

[45] T. Endo and L. O. Chua, "Chaos from phase-locked loops," *IEEE Trans. Circuits Syst.*, Vol. 35, pp. 987–1003, 1988.

[46] T. Endo, "A review of chaos and nonlinear dynamics in phase-locked loop," *J. Franklin Institute*, Vol. 331B, pp. 859–902, 1994.

[47] T. Endo and L. O. Chua, "Chaos in mutually coupled phase-locked loops," *IEEE Trans. Circuits Syst.*, Vol. 37, pp. 1183–1187, 1990.

[48] S. Hayes, G. Grebogi and E. Ott, "Communication with chaos," *Phys. Rev. Lett.*, Vol. 70, pp. 3031–3034, 1993.

[49] B. A. Harb and M. A. Harb, "Chaos and bifurcation in third-order phase-locked loop," *Chaos, Solitons, Fractals*, Vol. 19, pp. 667–672, 2004.

[50] B. C. Sarkar and S. Chakraborty, "Self-oscillations of a third order PLL in periodic and chaotic mode and its tracking in a response PLL," *Commun. Nonlinear Sci. Numer. Simul.*, Vol. 19, No. 3, pp. 738–749, 2014.

[51] V. S. Afraimovich, "Intrinsic bifurcations and crisis of attractors," *Nonlinear Waves. Structures and Bifurcations*, Nauka, Moscow, 1987 (in Russian).

[52] N. N. Bautin and E. A. Leontovich, *Dynamical Systems on the Plane*, Nauka, Moscow, 1976 (in Russian).

[53] V. V. Matrosov and D. V. Kasatkin, "Dynamic operating modes of coupled phase controlled oscillators," *J. Commun. Technol. Electron.*, Vol. 48. No. 6, pp. 637–645, 2003.

[54] V. V. Matrosov and M. V. Korzinova, "Simulation of nonlinear dynamic cascade phase-locked loops", *Radiophys. Quantum Electron.*, Vol. 36, No. 8, pp. 555–559, 1993.

[55] V. V. Matrosov and M. V. Korzinova, "Collective dynamics of cascade coupled phase systems," *Izv. VUZov, Prikladnaya Nelineynaya Dinamika*, Vol. 2, No. 2, pp. 10–16, 1994 (in Russian).

[56] V. V. Matrosov and M. V. Shalfeeva, "Influence of coupling parameters on the nonlinear dynamics of two cascade-coupled phase-locked loops," *Radiophys. Quantum Electron.*, Vol. 38, No. 3/4, pp. 180–184, 1995.

[57] V. V. Matrosov, "Some particularities of dynamical behavior of two cascade coupled phase locked loops," *Izv. VUZov, Prikladnaya Nelineynaya Dinamika*, Vol. 6, No. 6, pp. 52–61, 1997 (in Russian).

[58] N. N. Bautin, *Behavior of Dynamic Systems Near the Boundaries of the Stability Region*, Gostekhizdat, Moscow-Leningrad, 1949 (in Russian).

[59] V. D. Shalfeev, V. V. Matrosov and M. V. Korzinova, "Communications using cascade coupled phase-locked loops chaos," *Int. J. Bifurc. Chaos*, Vol. 9, No. 5, pp. 963–973, 1999.

[60] Yu. I. Neimark and P. S. Landa, *Stochastic and Chaotic Oscillations*, Kluwer Academic Publishers, Dordrecht, 1992.

[61] V. V. Matrosov, V. D. Shalfeev and D. V. Kasatkin, "Analysis of regions of chaotic oscillations in coupled phase systems," *Radiophys. Quantum Electron.*, Vol. 49, No. 5. pp. 406–414, 2006.

[62] K. G. Mishagin, V. V. Matrosov, V. D. Shalfeev and V. V. Shokhnin, "Experimental study of chaotic oscillations generated by an ensemble of cascade-coupled phase systems," *Tech. Phys. Lett.*, Vol. 31, No. 12, pp. 1052–1054, 2005.

[63] K. G. Mishagin, V. V. Matrosov, V. D. Shalfeev and V. V. Shokhnin "Generation of chaotic oscillations in the experimental scheme of two cascade-coupled phase systems," *J. Commun. Technol. Electron.*, Vol. 52, No. 10, pp. 1146–1152, 2007.

[64] V. V. Matrosov and D. V. Kasatkin "The analysis of excitation processes of chaotical oscillations in coupled PLLs," *Izv. VUZov, Prikladnaya Nelineynaya Dinamika*, Vol. 11, No. 4, pp. 31–43, 2003 (in Russian).

[65] V. V. Matrosov and D. V. Kasatkin "Dynamic properties of coupled phase-locked loop," *Izv. VUZov, Prikladnaya Nelineynaya Dinamika*, Vol. 12, No. 1, pp. 159–168, 2004 (in Russian).

[66] V. V. Matrosov and V. D. Shalfeev, "Excitation of chaotic oscillations in coupled PLLs," *Proc. 2nd IEEE Int. Conf. Circuits and Systems for Communications*, Moscow, Russia, 2004.

[67] V. V. Matrosov, "Dynamics of two parallel coupled phase systems," *Izv. VUZov, Prikladnaya Nelineynaya Dinamika*, Vol. 14, No. 1, pp. 25–37, 2006 (in Russian).

[68] L. M. Pecora and T. L. Carroll, "Synchronization in chaotic system," *Phys. Rev. Lett.*, Vol. 64. No. 8, pp. 821–824, 1990.

[69] N. F. Rulkov, M. M. Sushchik, L. S. Tsmring and H. D. I. Abarbanel, "Generalized synchronization of chaos in directionally coupled chaotic systems," *Phys. Rev. E*, Vol. 51, No. 2, pp. 980–994, 1995.

[70] L. Kocarev and U. Parlitz, "Generalized experimental synchronization, predictability and equivalence of unidirectionally coupled dynamical system," *Phys. Rev. Lett.*, Vol. 76, No. 11, pp. 1816–1819, 1996.

[71] M. G. Rosemblum, A. S. Pikovsky and J. Kurths, "From phase to lag synchronization in coupled chaotic oscillators," *Phys. Rev. Lett.*, Vol. 78, No. 22, pp. 4193–4196, 1997.

[72] S. Taherion and Y. C. Lai, "Observability of lag synchronization in coupled chaotic oscillators," *Phys. Rev. E*, Vol. 55, No. 6, pp. 6247–6250, 1999.

[73] V. S. Anishchenko, T. E. Vadivasova, D. E. Postnov and M. A. Safonova, "Synchronization of chaos," *Int. J. Bifurc. Chaos*, Vol. 2, No. 3, pp. 633–634, 1992.

[74] A. Pikovsky, M. Rosemblum and J. Kurths, *Synchronization: A Universal Concept in Nonlinear Sciences*, Cambridge University Press, 2002.

[75] A. S. Dmitriev, A. V. Kletsov, A. M. Laktyushkin, A. I. Panas and S. O. Starkov, "Ultrawideband wireless communications based on dynamic chaos," *J. Commun. Technol. Electron.*, Vol. 51, No. 10, p. 1126, 2006.

[76] V. D. Shalfeev, G. V. Osipov, A. K. Kozlov and A. R. Volkovskii, "Chaotic oscillations — generation, synchronization, controlling," *Zarubezhnaja Radio Electronika, Uspekhi Sovremennoi Radioelektroniki*, No. 10, pp. 27–49, 1997 (in Russian).

[77] G. V. Osipov, A. S. Pikovsky and J. Kurths, "Phase synchronization of chaotic rotators," *Phys. Rev. Lett.*, Vol. 88, pp. 0541021–0541024, 2002.

[78] B. Hu, G. V. Osipov, H.-L. Yang and J. Kurths, "Oscillatory and rotatory synchronization of chaotic autonomous phase systems," *Phys. Rev. E*, Vol. 67, pp. 0662161–0662169, 2003.

[79] V. D. Shalfeev and G. V. Osipov, "Chaotic phase synchronization of coupled PLLs," *Proc. NDES'97*, Moscow, Russia, pp. 139–144, 1997.

[80] V. D. Shalfeev and V. V. Matrosov, "Dynamical chaos in phase-locked loops," *Chaos in Circuits and Systems*, G. Chen and T. Ueta (eds.), World Scientific Publishing Co., Singapore, pp. 130–150, 2002.

[81] H. Dedieu, M. P. Kennedy and M. Hasler, "Chaos shift keying: modulation and demodulation of a chaotic carrier using self-synchronizing Chua's circuits," *IEEE Trans. Circuits Syst. II., Analog Digit. Signal Process.*, Vol. 40, No. 10, pp. 634–642, 1993.

[82] M. P. Kennedy, "Communication with chaos: state of the art and engineering challenges," *Proc. NDES'96*. Seville, Spain, pp. 1–8, 1996.

[83] T. Endo and L. O. Chua, "Synchronization of chaos in phase-locked loops," *IEEE Trans. Circuits and Syst.*, Vol. 38, pp. 1580–1588, 1991.

Index